Statistical Explanation and Statistical Relevance

Drawing by Richter; © 1970 The New Yorker Magazine, Inc.

Statistical Explanation & Statistical Relevance

Wesley C. Salmon

With Contributions by

Richard C. Jeffrey

and

James G. Greeno

UNIVERSITY OF PITTSBURGH PRESS

Richard C. Jeffrey, "Statistical Explanation vs. Statistical Inference," in *Essays in Honor of Carl G. Hempel,* ed. Nicholas Rescher, pp. 104–13. © 1969, D. Reidel Publishing Company, Dordrecht, Holland. Reprinted by permission of publisher and author.

Wesley C. Salmon, "Statistical Explanation," in *Nature and Function of Scientific Theories,* ed. Robert G. Colodny, pp. 173–231. Copyright © 1970, University of Pittsburgh Press.

James G. Greeno, "Evaluation of Statistical Hypotheses Using Information Transmitted," *Philosophy of Science,* 37 (June 1970), pp. 279–93. © 1970 by The Philosophy of Science Association. Reprinted by permission of publisher and author. This essay has been retitled "Explanation and Information."

Library of Congress Catalog Card Number 77–158191
ISBN 0–8229–5225–4

Henry M. Snyder & Co., Inc., London
Manufactured in the United States of America

To Merrilee

Contents

Preface

It was with a sense of intense excitement that I first learned of the work of Professors Richard Jeffrey and James Greeno, for it seemed to me that their ideas on statistical explanation fit beautifully with notions I was in the process of developing but had not yet published. I felt that we were all working toward a new conception of statistical explanation which constitutes a radical and much-needed departure from the received view. The three papers contained in this volume seem to me to constitute a trilogy of essays that go together harmoniously and supplement one another in fruitful ways. They are the first three publications in the elaboration of the *statistical-relevance* or S-R model of explanation.

I am deeply grateful to Jeffrey and Greeno for the illumination they have supplied and for permission to reprint their essays in this book. I should state quite emphatically, however, that Jeffrey and Greeno have not, in allowing me to reprint their articles, tacitly or explicitly endorsed my general views about this whole new approach to the problems of statistical explanation. In other words, they are not necessarily as enthusiastic about my ideas as I am about theirs.

I should also like to express my most sincere thanks to my colleagues J. A. Coffa, Ronald Giere, and Larry Wright, all of whom have read my essay with care and offered extremely insightful comments. I owe special gratitude to Professor Carl G. Hempel who, under circumstances that must have involved considerable personal inconvenience, read my paper with charity as well as critical acumen and offered a number of valuable observations, some of which are mentioned in the Postscript of this book. Further thanks are due to the D. Reidel Publishing Company and to *Philosophy of Science* for permission to reprint the articles by Jeffrey and Greeno, to the artist and *The New Yorker* for permission to reproduce the drawing on page ii, and to the University of Pittsburgh Press for permission (and encouragement) to reprint my essay. Finally, I should like to express my gratitude to the National Science Foundation for continued support of my research on this and other topics.

April 1971 W.C.S.

Statistical Explanation and Statistical Relevance

Introduction

Suppose that a small plane crashed upon takeoff from an airport near Denver on July 15, 1971. We ask why the crash occurred. Our interest in the event might be highly practical; the FAA, for instance, investigates such accidents in order to improve flying safety. When we know why occurrences of various types happen, we can often do something about controlling them. At the same time, our interests might be largely theoretical. To someone concerned with aerodynamics, the search for an explanation of the crash might be the result of sheer intellectual curiosity. In either case—these two motives are not mutually exclusive—we look to science for an explanation of the event; the explanation may be both practically useful and intellectually satisfying. Whether we pursue scientific investigations for the purpose of predicting and controlling our environment or for the sake of understanding the world in which we live, the search for explanations is at the heart of the endeavor.

As the inquiry gets underway, the investigators will establish a number of relevant facts, such as the type of aircraft involved, its mechanical condition, the load it was carrying, the length of the runway, and the height and location of the obstacle. In addition, they will take into account such relevant meteorological circumstances as the wind velocity, the atmospheric pressure, the temperature, and the relative humidity. Given all of these conditions, the investigators can determine the distance needed for takeoff to clear the offending barrier. Having ascertained that the wind was calm and having ruled out such causes as mechanical failure, they find that the atmospheric pressure was low (due to the high altitude of the airport) and that the day was hot and humid. Since the distance required for takeoff depends upon the density of the air—the smaller the density the greater the distance needed—and since density decreases as altitude, temperature, and humidity are increased, the conditions at the time and place of the accident resulted in an abnormally long takeoff distance. Under these circumstances the runway simply was not long enough. The pilot made the fatal error of failing to take these factors into account.

This example illustrates several important features of scientific explanation. We have a particular event—the crash of the airplane—for which an explanation is sought; it is known as the *explanandum* (that which is to be explained).[1] It is explained by invoking *general laws*—for example, that there is an inverse relation between the density of the air and the distance needed for takeoff—under which the explanandum can be subsumed. The explanandum is brought under the general laws by establishing such *initial conditions* as air temperature, atmospheric pressure, relative humidity, wind velocity, type of aircraft, and height and location of the obstacle. The general laws and the initial conditions together constitute the *explanans* (that which does the explaining). The explanatory facts which make up the explanans thus consist of *particular facts*—embodied in the initial conditions—and *general facts*—embodied in the general laws. In order to explain a particular occurrence, both types of facts are essential. The fact that the relative humidity was high on the day of the crash will not help to explain the crash unless we have a general law relating humidity to air density. Similarly, the general relationship between humidity and density is useless without the particular value of the humidity at the place and time of the crash, which is required to bring the generalization to bear on that particular occurrence. General laws are needed to relate particular explanatory facts to the explanandum; particular facts are needed to make the general laws applicable to the explanandum. For obvious reasons, explanations that conform to this pattern are called *covering-law explanations*, and the pattern itself is known as the *covering-law model* of explanation.

We have not, of course, offered a full explanation of the crash; in order to do so, it would be necessary to fill in many details that we have only sketched. Even if such details had been furnished, thereby providing a complete explanation of the crash, it might still seem reasonable to ask for explanations of one or more parts of the explanans. This does not mean that the explanation of the crash is incomplete, but only that there are *other* explanations that might be in order. For example, it may be fairly evident why air is less dense at greater altitudes than at lesser ones, but perhaps it is puzzling that humid air is less dense than dry air. This *general fact* can be explained by noting that, at specified values of pressure and temperature, a particular volume of gas contains approximately the same number of molecules regardless of the kinds of molecules composing it (Avogadro's law). Dry air contains mostly nitrogen (N_2) and oxygen (O_2) molecules, whereas humid air contains a significant proportion of water (H_2O) molecules. The molecular weights of N_2 and O_2 are 28 and 32 respectively, whereas the molecular weight of water vapor is 18. Although a wet washcloth is obviously much heavier than a dry one, a given volume of humid air is less massive, and hence less dense, than the same volume of dry air (at the same pressure and temperature).

We see, then, that general laws, as well as particular facts, are amenable to scientific explanation, and that the general law is explained by subsuming it under still broader laws. Thus, the general relation between the density of moist and dry air is explained in terms of still more general laws relating the density of a gas to its molecular composition. In this book we shall be concerned mainly with explanations of particular events rather than general laws, but it is important to remember that the general laws employed in such explanations are, themselves, capable of being explained by means of covering-law explanations.

In the context of aeronautical engineering, it is appropriate to regard the laws of physics as strict universal generalizations that hold without exception. Thus, the Bernoulli principle, which determines the lift of a wing, can be taken as an unexceptionable law relating the velocity of flow of a fluid (liquid or gas) to the pressure it exerts in a direction perpendicular to the direction of flow. From a more precise and theoretical standpoint, however, we must regard such laws as statistical generalizations that admit of overwhelmingly improbable exceptions. According to this more refined conception, the performance of an airplane attempting a takeoff is determined by the average behavior of exceedingly large numbers of molecules that collide with the propeller, wings, control surfaces, and other parts of the craft. If, for instance, an extremely large number of molecules of air near the obstacle, in the course of their purely random motions, had chanced to be moving upward at just the proper moment, they could have lifted the airplane over the obstacle, thus avoiding the accident. Such phenomena are actually observed for microscopic particles in Brownian motion, but for an object the size of an airplane such an occurrence is so incredibly improbable that for all practical purposes we can ignore its possibility. An occurrence of this type would be analogous to Jeffrey's example of a tire inflating spontaneously.[2]

The fact that certain laws are statistical in character, rather than strictly universal, obviously does not preclude their use in scientific explanations. If we ask why a particular ice cube melted, it would be adequate to point out that it was placed in lukewarm water and that an ice cube in these circumstances will very, *very* probably absorb heat from the surrounding water. It is not physically impossible, however, for the ice cube to give up some of its heat to the water, increasing rather than decreasing the temperature difference between them. Similarly, in strictest rigor the FAA investigators ought to say that an airplane of specified characteristics will very probably follow a particular flight path under particular meteorological conditions. This probabilistic construal of the explanation does not deprive it of its explanatory power. It is still a covering-law type of explanation, but the law is statistical rather than universal.[3]

It is evident from the foregoing remarks that statistical generalizations can be explained in much the same fashion as universal generalizations, namely, by subsumption under broader generalizations. Thus, for instance, the fact that two bodies of unequal temperature will probably exchange heat when brought into thermal contact, the warmer losing heat to the cooler until a uniform temperature is attained throughout, is explained by statistical generalizations concerning the exchange of kinetic energy among colliding molecules. In similar fashion, an explanation of the Bernoulli principle *as a statistical generalization* can also be grounded in the theory of the statistical behavior of gas molecules, inasmuch as pressure exerted on an airfoil by a gas is understood in terms of the averages of vast numbers of collisions of gas molecules with the surface.

The covering-law model of explanation demands, as we have seen, that every explanans contain at least one general law. The laws invoked in the explanation may be either universal or statistical generalizations. Whichever type of law is employed in the explanans, there seems to be a hierarchy of explanations, beginning at the lowest level with explanations of particular events and progressing upward through explanations of general laws of greater and greater scope. In fact, it has often been suggested that particular observable events, such as the airplane crash, can be explained through the use of *empirical laws*, whereas empirical laws are in turn explained by means of *theories*. This distinction between empirical laws and theories depends upon a rough distinction between observables and unobservables. An empirical law embodies general relations among more or less directly observable things and their more or less directly observable properties, whereas theories make reference to unobservables. Airplanes, runways, cylinders of air, and barometers, for instance, are observable entities, whereas molecules of nitrogen and oxygen are not. Likewise, the atmospheric pressure, the weight of an airplane, the wind velocity, and the length of the runway are observable properties of observable things, whereas the number of molecules in a container of gas and the kinetic energy of a single molecule are not. Consequently, the inverse relation between relative humidity and takeoff distance is an *empirical law* which can be used to explain a particular event, the airplane crash, but the generalizations which involve weights of individual molecules and numbers of molecules in a given volume of gas are highly theoretical in character. It was by means of such *theoretical laws* that we explained the foregoing *empirical law,* and this seems typical of the way in which empirical laws are theoretically explained. Theoretical laws, themselves, are often explained in terms of more general and fundamental theories. In this book we shall confine attention chiefly to the explanations of particular events which stand at the bottom of the hierarchy. Once the explanations of the lowest level are understood, we can perhaps hope to move upward and cope with higher-level explanations.

Although the foregoing examples illustrate much of what is involved in scientific explanations, there remains the surprisingly difficult *philosophical* task of providing a general characterization of the logical structure of scientific explanation. This is an ancient problem; as Jeffrey points out in his essay, it goes back at least to Aristotle. Nevertheless, it has been the object of vigorous and fruitful investigation for the past quarter century, due largely to the work of Carl G. Hempel. Beginning with a classic article in 1948, Hempel has elaborated a remarkably appealing and admirably precise theory of the nature of scientific explanation.[4] Although Hempel's account has not been universally accepted by philosophers of science, it has been extremely influential, and it constitutes the closest thing we have to a received view. An extensive literature has grown up around it. For these reasons Hempel's account serves as the very best point of departure for any contemporary discussion of scientific explanation.

Hempel has advanced two basic models of scientific explanation of particular events. The first is the so-called *deductive-nomological* (D-N) pattern; its explanans consists of *universal* laws and initial conditions, from which the occurrence of the explanandum follows deductively. The deductive-nomological explanation *is* a valid *deductive argument;* the statements asserting the initial conditions and universal laws of the explanans are the premises, and the assertion of the explanandum is the conclusion. The second is the so-called *inductive-statistical* (I-S) pattern. The general law in this type of explanation is a *statistical* generalization, and the explanation *is* an *inductive argument.* The statements of initial conditions and statistical laws in the explanans do not necessitate the explanandum, but they do confer high inductive probability upon it. Harking back to our example of the airplane crash for illustration, we can see that it would be possible in principle, though enormously complicated in fact, to construct a valid deduction of the occurrence from strictly universal laws and sufficiently detailed specifications of initial conditions. If, however, we decide for the sake of greater rigor to treat the laws of aerodynamics as statistical generalizations, we could presumably construct an inductive argument according to which the crash was overwhelmingly probable. The logical details of these two models of scientific explanation are presented, along with simple examples of their application, in section 1 of my essay, "Statistical Explanation." For the moment, it is sufficient to observe that *explanations of each type are arguments,* deductive or inductive, *showing that the event to be explained*—the explanandum—*was to be expected by virtue of the explanatory facts set forth in the explanans.*

Deductive validity is an all-or-nothing affair; deductive arguments are either valid or invalid, and there are no degrees of validity. Inductive support, by contrast, does admit of degrees. The premises of an inductive argument may lend more or less weight to the conclusion, and one may speak of degrees of strength of inductive inferences. According to Hempel's view, an inductive-statistical explanation, therefore, has a degree of strength which he designates as the *inductive probability* conferred upon the explanandum by the explanans.

Deductive logic is a highly developed discipline, and any questions about deductive validity that are likely to arise in the course of the discussion of scientific explanation can be settled rather straightforwardly. Inductive logic, by contrast, is in a rather primitive state of development, and many fundamental problems about inductive-statistical explanation cannot easily be settled by referring to a canonical system of inductive logic. If, for example, we were to ask for further elucidation of the concept of inductive probability, it would be reasonable to go to Rudolf Carnap's theory of confirmation, the most rigorous and extensive system of inductive probability available at present.[5] When we look at Carnap's inductive logic, we discover a shocking fact: in that system of inductive logic (the one to which Hempel explicitly refers in connection with the concept of inductive probability), *there is no such thing as inductive inference* in the sense required for Hempel's account of inductive-statistical explanation! In Carnap's inductive logic there are no inductive arguments consisting of premises and conclusion, which allow you to affirm the conclusion (with some degree of probability) if you are prepared to assert the premises. On this view, inductive logic is strongly disanalogous to deductive logic, even to the extent of proscribing inference entirely.

This is not the place to go into Carnap's reasons for denying the possibility of inductive inferences or to argue the merit of his arguments. Nor do I mean to suggest that Carnap's system of inductive logic is the only possible one. But an important heuristic point is in order. Since Carnap's account of inductive probability is the most prominent and best-developed theory available, there does seem to be good reason to wonder whether statistical explanations are arguments at all. This is precisely the revolutionary move made by Jeffrey in his essay, "Statistical Explanation vs. Statistical Inference," and it is the first step in developing an alternative to Hempel's models. Jeffrey concludes, roughly speaking, that the statistical explanation of an event exhibits that event as the result of a stochastic process from which such events arise with some probability whose degree may be high, middling, or even very low. Exhibition of such a process does not constitute an argument at all, let alone an argument to the effect that the explanandum was to be expected by virtue of its high probability.

If we take seriously the suggestion that some events have low probabilities, a further reason emerges for advancing the alternative to Hempel's theory of explanation. Were we to insist, with Hempel, that a statistical explanation must embody a high probability, then events that are intrinsically improbable, even though they sometimes occur, would consequently defy all explanation. For example, in the light of current physical theory, the spontaneous radioactive decay of a uranium atom may be due to an alpha particle tunneling through the potential barrier of the nucleus. As the alpha particle bombards this nuclear wall, there is a probability of the order of 10^{-38} that it will escape, and there are no further relevant factors to determine in which instance it tunnels out. On Hempel's view such low probability events are in principle incapable of being explained, but on the alternative account quantum mechanics provides an explanation by furnishing all of the facts relevant to their occurrence.

There is, of course, a very strong temptation to maintain that there must be *some* reason why the alpha particle gets through on one occasion, when it fails on so many others, but the reasons are not known at present. This view involves an a priori commitment to determinism—that is, to the doctrine that every event that happens is completely determined by previous causes. On this view our reliance upon probabilities is simply a reflection of our ignorance; further investigation will reveal the unknown causes and enable us to give a full (deductive-nomological) explanation of the event in question. This position seems to me untenable. I do not mean to argue that present physical theory is complete and correct but, rather, that there is no reason to make an a priori decision as to the nature of further physical theories. Perhaps, in the future, improved theories will provide a deterministic account of events that current theory regards as causally undetermined—but perhaps they will not. We should be prepared for the possibility that the indeterministic character of physical theory is correct and that there are events which are intrinisically improbable, not merely improbable in relation to our present incomplete knowledge. In that case, we need an account of statistical explanation that will characterize the explanation of events in terms of statistical laws. It seems desirable for a theory of explanation to admit the possibility of events that are intrinsically undetermined, indeed, events whose intrinsic probabilities are low, without denying the possibility of explaining them. To deny that any events are undetermined seems to involve an unwarranted a priori commitment to determinism; to say that only events with high probability can be explained involves, among other disadvantages, the acute embarrassment of trying to say in some nonarbitrary way how high is high enough.

Jeffrey, Greeno, and I all agree that statistical explanations need not be regarded as inductive arguments, and we agree that a high probability is not required for a correct statistical explanation. If high probability is not the desideratum, what can we offer as a substitute? The answer is *statistical relevance*. This is the view I have tired to elaborate in detail in my essay "Statistical Explanation." To see why statistical relevance is the key concept, consider the case of a person who experiences relief from a neurotic symptom while (or shortly after) undergoing psychotherapy. Does the psychotherapeutic treatment explain the remission of the symptom? The answer to this question depends not only upon the probability of the abatement of symptoms during (or shortly after) therapy; rather, it depends upon the relation between the remission rate for patients undergoing a particular type of treatment and the spontaneous-remission rate. Even if the probability of the remission of symptoms for patients in psychotherapy were very high, that would have no explanatory value if the spontaneous-remission rate were equally high. At the same time, even if the recovery rate for patients were quite low, but still higher than the spontaneous-remission rate, the fact that the individual had submitted to treatment would have some explanatory force in relation to his psychic improvement.

To say that a certain factor is *statistically relevant* to the occurrence of an event means, roughly, that *it makes a difference to the probability of that occurrence*—that is, the probability of the event is different in the presence of that factor than in its absence. This relation of statistical relevance, and its importance to the concept of statistical explanation, is illustrated by a recent development. In my essay, "Statistical Explanation," I introduced as an example the use of vitamin C as a cure for the common cold. At that time I was unaware of Dr. Linus Pauling's views on the efficacy of vitamin C for that purpose, and I quoted what then seemed fairly reliable evidence that the use of vitamin C is statistically irrelevant to recovery from a cold.[6] If Dr. Pauling is right, the use of vitamin C is relevant to recovery, and my previous factual information was incorrect. Clearly, however, the vital question is not "How probable is recovery from a cold if one takes sufficient vitamin C?" but rather "How does the probability for recovery differ between users and non-users of vitamin C?"

The foregoing considerations allow us to distinguish quite succinctly between Hempel's view and the alternative. Let us dub the alternative account "the *statistical-relevance* model" or "S-R model" for short. The term "inductive" is deliberately omitted from the title to emphasize that S-R explanations are not arguments or inferences of any sort. The two models can be characterized as follows:

> I-S model (Hempel): an explanation is an *argument* that renders the explanandum *highly probable.*
>
> S-R model (Jeffrey, Salmon, Greeno): an explanation is an *assembly of facts statistically relevant* to the explanandum, *regardless of the degree of probability* that results.

It is evident that an explanation can satisfy Hempel's high-probability requirement without satisfying the relevance requirement and that the relevance requirement can be fulfilled in the absence of high probability. The fact that high probability is neither necessary nor sufficient for statistical relevance indicates that the difference between Hempel's I-S model and our S-R model is fundamental. In my essay I attempt to explain in detail how one goes about assembling sets of conditions relevant to the occurrence of an event—indeed, even *complete* sets of relevant conditions—and to offer further justification for characterizing statistical explananation in that way. Furthermore, I offer counterexamples—

such as the man who takes birth control pills and avoids becoming pregnant—to show that even deductive-nomological explanations can fail on account of lack of relevance. Richter's excellent drawing which serves as a frontispiece illustrates the same point. We are led to the suggestion that explanations embodying universal generalizations are simply a limiting case of S-R explanation, subject to the same kinds of relevance requirements.

When we look at the airplane crash from the standpoint of relevance, we may start by asking why a brand X airplane in good mechanical condition carrying a reasonable load should fail to clear an obstacle when similar craft with similar loads had often taken off successfully from runways no longer than this one. Whether we are construing the laws as universal or statistical, we want to find *relevant* conditions to account for the crash. The answer, we find, is the air density at the time and place of the crash. It appears that the pilot had made the all-too-common error of forgetting the relevance of altitude, temperature, and humidity to the distance required for takeoff.

Having argued rather adamantly that events with low probabilities are amenable to scientific explanation, I must confess to a feeling of queasiness in saying that an event is explained when we have shown that according to all relevant factors, its occurrence is overwhelmingly improbable. I am somewhat inclined to attribute this feeling to intuitions that have been well nurtured on more than two decades of exposure to Hempel's very persuasive writings, and to say that we simply have to retrain our intuitions. Greeno has a different, and I suspect better, way of handling this matter. Given that a theory (i.e., a collection of statistical laws) has to explain the occurrence and nonoccurrence of many different types of events, and that factors relevant to the nonoccurrence of the event seem to have a place in the explanation,[7] Greeno suggests that we evaluate the overall explanatory power of a theory to explain all of the kinds of events it purports to explain, rather than attempting to evaluate the goodness of a particular explanation of a particular event. One of the attractive features of Greeno's essay, "Explanation and Information," is that it provides an appealing method of assessing at least one aspect of the explanatory value of a theory. This result is achieved by application of some concepts of information theory.

When we ask what good it is to have an S-R explanation, it is satisfying to be able to say that the invocation of an explanation increases our information. Indeed, information theory even provides a quantitative measure of the amount of increase. Perhaps there are other desiderata for explanatory theories, but increase of information is a nice one. Looking at the measure in some detail, we shall see that the addition of information accrues as a result of the fact that the explanation provides appropriate relevance relations.

Consider a simple example. Suppose that the population of Centerville, U.S.A., is equally divided between Democrats and Republicans. Let us call the partition of the population in terms of political affiliation $\{M\}$ (to be thought of as explanandum), and let

$$M_1 = D;\ M_2 = R; \text{ where } \{M\} = \{D, R\}.$$

According to our assumption,

$$p_1 = P(M_1) = P(D) = 1/2;$$
$$p_2 = P(M_2) = P(R) = 1/2.$$

This partition involves the greatest possible degree of uncertainty for a partition into two subclasses, for knowing that a person is a resident of Centerville tells us nothing about whether he is a Republican or a Democrat. In information theory this uncertainty is measured by

$$H(M) = \sum_i -p_i log_2 p_i = 1,$$

where it is sometimes called, with enormous potentiality for confusion, the "information."[8] The bifurcation into two equally probable subsets provides the unit of uncertainty (or information) known as the "bit." Notice that the uncertainty achieves its maximum value of 1 when $p_1 = p_2$, and it drops to its minimum of zero when either p_1 or p_2 assumes the value of 1. If all residents of Centerville were Republicans, there would be no uncertainty whatever about their party affiliation.

Now suppose, moreover, that Centerville is split by a set of railroad tracks that run north to south through the town, so that half of the residents live to the east and half live to the west of the tracks. Here we have another partition; let us call it $\{S\}$ (to be thought of as explanans), and let

$$S_1 = E;\ S_2 = W; \text{ where } \{S\} = \{E, W\}.$$

According to our second assumption

$$p'_1 = P(S_1) = P(E) = 1/2;$$
$$p'_2 = P(S_2) = P(W) = 1/2.$$

Again, the uncertainty is maximal for such a partition:

$$H(S) = 1.$$

The aggregate uncertainty of the two partitions is their sum,

$$H(M) + H(S) = 2.$$

The important question about these two partitions concerns their mutual independence. Let us call the unconditional probabilities $P(S_i)$ and $P(M_j)$ given above the "marginal probabilities." Let us then introduce the conditional probabilities $P(M_j,S_i) = p_{ij}$ from the members of the partition $\{S\}$ to the members of the partition $\{M\}$.[9] By definition, the partition $\{M\}$ is independent of the partition $\{S\}$ if the conditional probabilities are equal to the respective marginal probabilities:

$$P(M_j, S_i) = P(M_j) \quad \text{or} \quad p_{ij} = p_j \text{ for every } i, j.$$

Intuitively we want to say that the conditional probabilities contribute no further information if the two partitions are statistically independent of one another, but that they can contribute positive information (reduction of uncertainty) if there is a statistical dependency between them. The general idea is this: if the probability of being a Democrat varies, depending upon the side of the tracks on which the resident lives, then knowledge of his place of residence reduces the uncertainty about his party affiliation. If, however, $P(D,E) = P(D,W) = P(D)$, then knowledge of place of residence does not provide any information relevant to party affiliation.

In information theory, the quantitative measure of reduction of uncertainty—that is, the "information transmitted" by the theory T—is given by

$$I_T = H(M) + H(S) - H(S \times M)$$

where

$$H(S \times M) = \sum_i \sum_j -p_i p_{ij} log_2 p_i p_{ij}.$$

As Greeno shows,

$$H(S \times M) = H(M) + H(S)$$

whenever the conditional probabilities equal the corresponding marginal probabilities (i.e., $p_{ij} = p_j$). Thus, if the partitions are independent, the reduction in uncertainty, or the increase of information, is zero. This, of course agrees with our intuitions. It also means that a partition $\{S\}$ that is statistically irrelevant to a partition $\{M\}$ cannot have any explanatory value with respect to it.

Suppose, however, that the two partitions are not independent and that place of residence is relevant to party affiliation. In particular, let us say that 3/4 of the people to the east of the tracks are Democrats, whereas 3/4 of those on the west side are Republicans:

$$P(D, E) = P(M_1, S_1) = p_{11} = 3/4;$$
$$P(R, E) = P(M_2, S_1) = p_{12} = 1/4;$$
$$P(D, W) = P(M_1, S_2) = p_{21} = 1/4;$$
$$P(R, W) = P(M_2, S_2) = p_{22} = 3/4.$$

Then,

$$H(S \times M) = -2(3/8 \log_2 3/8 + 1/8 \log_2 1/8) \simeq 1.81$$

and

$$I_T \simeq 0.19.$$

This quantity represents our increase of information by virtue of the conditional probabilities.

Greeno shows that the increase in information is maximal when all of the conditional probabilities are either zero or one. This corresponds to the situation in which deductive-nomological explanation is possible. If, however, the marginal probabilities in the original explanandum partition are also either zero or one, that maximum represents no gain in information. This situation corresponds to the case in which deductive-nomological explanation becomes vacuous through failure of relevance conditions, as in the example of the man who takes birth control pills. These considerations show quite clearly that the measure of explanatory value introduced by Greeno is a statistical-relevance measure. An explanatory theory can, on his view, have explanatory value even though it assigns low probabilities to explanandum events, and it may fail to have explanatory value even if it assigns high probabilities to explanadum events.

In my essay, "Statistical Explanation" (section 6), I approach the increase in information resulting from a relevant partition in a different way. Suppose we select a particular resident of Centerville, and ask for the probability that he is a Democrat. We would be ill-advised to accept the value 1/2—say as the basis for a 50–50 bet— because the reference class of residents is not homogeneous with respect to party affiliation. We should look instead at his residence; if he lives on the east side of the tracks, assign the value 3/4, and if he lives on the west side, the value should be 1/4. Let us look at the matter quantitatively by associating the (true) value 1 with each Democrat and the (true) value 0 with each Republican. If we assign the value 1/2 to each person, the error (deviation from the "true" value) in each case is $\pm$ 1/2, and (squaring to make everything positive) the squared error is 1/4 for each individual. Obviously, the mean-squared error for all N residents is 1/4. Suppose, instead, that we assign the value 3/4 to each resident east of the tracks, and the value 1/4 to each resident of the west side. The number of residents on each side is $N/2$; $3/4 \times N/2$ of the east siders are Democrats, while $1/4 \times N/2$ are Republicans. The converse situation obtains on the west side. If we assign the value 3/4 to someone who is a Democrat, the error is 1/4 and the squared error is 1/16. If we assign the value 3/4 to a Republican, the error is 3/4, and the squared error is 9/16. The cumulative squared error for all residents of the east side is

$$3/4 \times N/2 \times 1/16 + 1/4 \times N/2 \times 9/16 = 12N/128.$$

The same cumulative squared error occurs if we assign the value 1/4 to each west-sider. The total cumulative squared error is $24N/128$, and the mean-squared error is 3/16, which is less than 1/4, the mean-squared error that results from ignoring the relevant partition in terms of place of residence. We see once more how the increase in information due to a relevant partition of the reference class translates into a numerical measure—in this case, one that is rather obviously related to predictive success.

Greeno has chosen to explicate S-R explanation in terms of information transmitted, whereas I have chosen to explicate it in terms of the homogeneity of the reference class. As we have seen, each approach leads to a quantitative measure, as well as a qualitative characterization. Qualitatively, the explications seem to coincide, for they agree that the essence of explanation is in the relevance relations expressed by the conditional probabilities that relate the explanans partition to the explanandum partition. Both of these treatments provide a straightforward answer to a previously recalcitrant problem concerning the utility of scientific explanations. On the present view some of the values of an S-R explanation are the increase of information, the decrease in uncertainty, and the increase in predictive success provided by genuine explanations. For starters, they seem to supply a reasonable motivation for looking more closely at the structure of S-R explanation.

NOTES

1. For the moment I am being deliberately ambiguous about the nature of the explanandum—whether it is the event or the statement that the event occurs. This technicality will be treated herein in sec. 1 of my essay "Statistical Explanation." Similar remarks apply to the explanans.

2. See his essay, "Statistical Explanation vs. Statistical Inference," reprinted in this volume.
3. I trust no confusion will arise from the fact that I use the same example to illustrate both types of explanation. If our present physical theories are more or less adequate, the statistical explanation is the correct one, but the explanation via universal laws is so nearly correct that it suffices for practically every purpose.
4. Carl G. Hempel and Paul Oppenheim, "Studies in the Logic of Explanation," *Philosophy of Science*, XV, pp. 135–75.
5. The most extensive exposition is found in his *Logical Foundations of Probability* (Chicago: University of Chicago Press, 1950).
6. Linus Pauling, *Vitamin C and the Common Cold* (San Francisco: W. H. Freeman and Co., 1970).
7. I argue this point at length in "Statistical Explanation," sec. 9.
8. Use of logarithms to the base 2 is customary in information theory, but Greeno uses natural logarithms in his essay "Explanation and Information." Nothing important hinges on this choice; the use of base 2 merely yields a convenient unit of information.
9. Here I write conditional probabilities in conformity to customary usage and Greeno's notation, not in Reichenbach's rather idiosyncratic reverse notation which I have employed in "Statistical Explanation."

Statistical Explanation vs. Statistical Inference

RICHARD C. JEFFREY

University of Pennsylvania

Hempel is not the first philosopher to have held that causal explanations are deductive inferences of a special sort: in the *Posterior Analytics*[1] Aristotle distinguishes a special sort of deductive inference – the demonstrative syllogism – in these terms:

By demonstration I mean a syllogism productive of scientific knowledge, a syllogism, that is, the grasp of which is *eo ipso* such knowledge.

He then lays down defining conditions for this special sort of inference:

... the premisses of demonstrated knowledge must be true, primary, immediate, better known than and prior to the conclusion, which is further related to them as effect to cause.

And he remarks,

Syllogism there may indeed be without these conditions, but such syllogism, not being productive of scientific knowledge, will not be demonstration.

Now we can fault this account on various grounds, but so can we fault contemporary accounts. We must give the old man credit; as he says at the end of the *Organon* (at the end of *De Sophisticis Elenchis*), his was the first book on logic; and he concludes,

... there must remain for all of you, or for our students, the task of extending us your pardon for the shortcomings of the inquiry, and for the discoveries thereof your warm thanks.

The affinities between the Hempelian and Aristotelian accounts of explanation may be obscured by differences in terminology. Thus, Aristotle speaks of syllogism, Hempel of deductive inference; and Aristotle speaks of knowledge, Hempel of explanation. But remember that 'syllogism' was Aristotle's general term for deductive inference – and do not tax him with what we now know to be his overly narrow view of the forms that such inferences can take. As to knowledge vs. explanation, Aristotle says

We suppose ourselves to have unqualified scientific knowledge of a thing, as opposed to knowing it in the accidental way in which the sophist knows, when we think t hat we know the cause on which the fact depends ....

It is precisely this sort of understanding that is conveyed by causal explanation – 'the grasp of which is *eo ipso* such knowledge'. Then I take it that Aristotle's *demonstrated knowledge* – knowledge, as he says later, not simply of the fact but of the *reasoned fact* – is the psychological correlate of Hempel's causal explanation as it is of Aristotle's demonstrative syllogism.

Of course, I have been overemphasizing the similarities between the Aristotelian and Hempelian accounts of scientific understanding; and of course, my point is not to advocate a back-to-Aristotle movement. Still less do I seek to belittle Hempel's work on explanation; the point is rather to appreciate that work as continuous with the philosophical enterprise that started in Athens 2400 years ago.

My real concern here is not with the history of nomological-deductive accounts of explanation – interesting as that may be – nor is it with the nomological-deductive view of causal explanation itself – satisfactory as *that* may be. My concern is with statistical explanation, and I have begun by recalling a general feature of nomological-deductive explanation because I want to show that statistical explanations often lack that feature. The particular feature is that of *being an inference*. On Aristotle's account as on Hempel's, causal explanations are deductive inferences; and it seems plausible to say that for their part, statistical explanations are statistical inferences – or, if you prefer, inductive or probabilistic inferences. But I shall argue that only in certain cases is it plausible to view statistical explanations as statistical inferences.[2]

The really beautiful cases of statistical explanation are the statistical mechanical ones in which the explained occurrence can be shown to have probability so close to 1 as to 'make no odds' in any gamble or other decision problem. I am thinking here of such cases as the explanation of why a flat tire does not spontaneously inflate: it is *possible* that the random movements of the surrounding air molecules might, for a few seconds, be such as to constitute a jet of air through the puncture powerful enough to pump the tire up – possible, but so improbable as to make it a *practical certainty* that no such jet will be forthcoming. (A proposition is a practical certainty if its probability is so high as to allow us to reason, in *any* decision problem, as if its probability were 1.) It is in these beautiful, extreme cases that the view of statistical explanations as statistical inferences is no less plausible than the view of causal explanations as deductive inferences.[3]

But there is a certain obliqueness about these inferential explanations, whether the strength of the inference be deductive certainty or practical certainty. I mean, one explains in order to impart knowledge of *how* or *why* the explained phenomenon takes place, but the explanation itself (in these cases) takes the form of a proof that the phenomenon *does* take place. The explanation is just the sort of thing one might produce in order to prove *that* the explained phenomenon takes place; this is the famous parallelism between explanation and prediction which I think breaks down for statistical explanations that impart less than practical certainty to the phenomenon explained.

I say that causal explanations (and the 'beautiful' kind of statistical explanations) are oblique: they explain *how* or *why* by demonstrating *that* in a special way. (Compare Aristotle's conception of knowledge of the *reasoned* fact.) Of course, not every demonstration *that* serves to explain *why*, as Louis MacNeice's rhyme testifies:

> The glass is falling hour by hour, the glass will fall forever.
> But if you break the bloody glass you won't hold up the weather.[4]

The falling of the glass does not cause the weather to turn bad; it is rather an effect of something that the bad weather is also an effect of. Suppose it were true that *whenever* the glass falls, the weather turns bad. Then the inference

The glass is falling.
Whenever the glass falls the weather turns bad.

∴ The weather will turn bad.

would support a prediction *that* the weather will turn bad without explaining *why* the weather turns bad. On the other hand, the inference

The glass is falling.
Whenever the glass is falling the atmospheric pressure is falling.
Whenever the atmospheric pressure is falling the weather turns bad.

∴ The weather will turn bad.

would serve both to predict and explain the weather: it gives us knowledge not merely of the fact, but of the reasoned fact, for it proves that the fact is a fact by citing causes and not mere symptoms. In general, where the inference meets certain conditions, one of which is that a causal law appear among the premisses, deductively grounded prediction will double as explanation. In these cases, knowledge *that* is imparted in such a way as to provide knowledge *why* as well. I mention this in order to suggest that this way of telling why obliquely, by demonstrating that, may not be the only way or the best way of telling why: the fact that we can sometimes explain by inferring is not a very strong reason to suppose that we can *only* explain by inferring. Perhaps, then, we should have another look at causal explanations, to see whether they *can* always be cast in the Aristotelian-Hempelian mould; but that is for another occasion. Here I want to have a look at statistical explanations, where I think it can pretty readily be seen that many of them simply cannot be represented as statistical inferences which show *why* obliquely, by demonstrating *that*.

Consider the following set of coin-tossing examples. A fair coin is tossed n times ($n=1, 2, 3, \ldots$), and at least one head turns up. Here the Hempelian statistical inference would have as its premiss a statistical law indicating that the probability of a head on any toss is $\frac{1}{2}$ and that distinct tosses are statistically independent, and would have as its conclusion the statement $H_1 \vee H_2 \vee \cdots \vee H_n$ that at least one of the first n tosses yields a head. The premiss – the statistical law – might be specified in various equivalent ways, e.g., by saying that the probability of the conjunction of any k distinct H's is $\frac{1}{2}^k$. From this specification one can deduce, via the laws of the elementary probability calculus, the statistical probability of any truth functional compound of the H's, e.g., the statistical probability of the conclusion of the statistical inference which Hempel represents as follows:

$$\frac{\text{If } i_1, \ldots, i_k \text{ are all distinct, then } p(H_{i_1} \,\&\cdots\&\, H_{i_k}) = \frac{1}{2}^k}{H_1 \vee H_2 \vee \cdots \vee H_n} \; [1 - 2^{-n}]$$

The bracketed number at the right represents the inductive probability – the degree of confirmation – of the conclusion, conditionally upon the premiss. If we represent the premiss by '*P*' and the conclusion by '*C*' then the foregoing array is to be viewed as a typographical variant of the statement

$$c(C, P) = 1 - 2^{-n}$$

which gives $1 - 2^{-n}$ as the inductive probability or degree of confirmation of the conclusion on the premiss.[5]

Hempel would hold that with $n=1$ the foregoing inductive inference is too weak to serve as an explanation of the fact that there was at least one head on the first n tosses: with $n=1$ the bracketed number – what Hempel calls *the strength of the explanation* – is only $\frac{1}{2}$. At the other extreme – for such large values of n as 100 – even I will admit that the inference will serve as an explanation, for a gap of $\frac{1}{2}^{100}$ between the probability of the conclusion and 1 is so small as to make no odds in any deliberation. But Hempel might hold that even with $n=10$ we have an explanation: the bracketed 'strength' of the inference is then 1023/1024 or a bit over .999.

Now I agree that with $n=10$ the strength of the inference is great enough to justify giving long odds, on the order of 1000:1, on there being at least one head. The number 1023/1024 is a good measure of the proper strength of our expectation of the fact; but I deny that it is a good measure of the quality or strength of the explanation which the inference gives us. Indeed, I think it misleading to think of the statistical inference as being an explanation at all. The explanation how or why there was at least one head, I should think, would be given by specifying the statistical probability function p that governs the process, e.g., as in the premiss of the inference, above. Perhaps it *is* part of the explanation to point out (as in the brackets above) that the probability of the explained phenomenon is then $1-(\frac{1}{2}^n)$, *but the strength of the explanation would be no greater with* $n=10$ *than with* $n=1$! To say, in the case $n=1$, that the statistical probability was $\frac{1}{2}$, is as much of an explanation as one can give. Indeed, this is a causal explanation in a sense in which a proof that there is at least one head can never be, for to say that the statistical probability is $\frac{1}{2}$ is to say *directly* something about cause which is said only obliquely, if at all, in a proof that there is at least one head. What is being said about cause, when we say that the statistical probability is $1-(\frac{1}{2}^n)$ that there will be at least one head, is that this effect is 'caused' by a chance or random process: we are saying directly that the usual sort of causality is absent.

To see this point it is helpful to consider what we should say if the improbable happened, say in the case $n=2$, and there were two tails. The strength of the 'explanation' of $-H_1$ & $-H_2$ would then be $\frac{1}{4}$ on Hempel's account, but we still have as complete an understanding of the why and the how of the outcome as we would have had if a head had turned up and the outcome had been the more probable one.

Moral: The strength of a statistical explanation (except in Hempel's technical sense) is not given by the degree of confirmation that the premisses bestow on the conclusion in the corresponding Hempelian inductive inference. Similarly for $n=10$: it is possible, although highly unlikely, that there will be ten tails, and if this happens we shall know all there is to know about the why of it and the how, when we know that the process which yielded the ten tails is a random one and when we know the probabilistic *law* governing the process. The knowledge that the process was random answers the question, 'Why?' – the answer is, 'By chance'. Knowledge of the probabilistic law governing the process answers the question 'How' – the answer is, 'Improbably, as a product of such-and-such a stochastic process'. Note that knowledge of the probabilistic law of the process makes it at most a matter of calculation to find that the statistical probability of the phenomenon – ten tails – is 1/1024. To know the probabilistic law is to know, among other things, that the actual outcome was hardly to have been expected.

Here knowledge *why* splits clearly away from knowledge *that*. Indeed in the statistical case I find it strained to speak of knowledge *why* the outcome is such-and-such. I could rather speak of *understanding the process* which had the outcome, for the explanation is basically the same no matter what the outcome: it consists of a statement that the process was a stochastic one, following such-and-such a law. (One may gloss this statement by pointing out that the actual outcome had such-and-such a probability, given the law of the process; but this gloss is not the heart of the explanation.)

Put the matter in this way: Aristotle's *knowledge of the reasoned fact* just will not do in the statistical case, for where a statistical explanation is appropriate, there's *no* reason for the fact: it came about by chance. Nor is the situation changed when (as usual) the probable happens. Because such cases are usual, we can usually give a statistical *inference* of strength $\frac{1}{2}$ or more then we can give a statistical explanation; and this, I take it, is why it is easy to mistake the inference for an explanation.

That the inference is *not* an explanation is shown, I think, by the fact that even when the improbable chances to happen, we give the same sort of account: the happening was the product of a *stochastic* process following such-and-such a probabilistic law. And we gloss this by pointing out that in this case the unexpected happened. My point is that it is no less a gloss, and no more essentially a part of the explanation, when we point out in the more usual cases that the *expected* happened.

Let us conclude by examining some cases in which explanation can be both causal and statistical. *Why was my first child a boy*? There are two sorts of answers: the statistical one ('There is no damned reason – it was pure chance'), and the causal ('Because the germ cell I contributed to the zygote which developed into the child was of the *Y* genotype'). Both answers are right; they supplement each other. The full-blown causal explanation might look like this:

> The sperm which united with an ovum to form the zygote out of which my first child developed had the *Y* genotype. Whenever the sperm which unites with an ovum to form the zygote out of which a child develops has the *Y* genotype, the child is a boy.
>
> ---
>
> My first child was a boy.

Here the first premiss could not have been known to be true before the birth, and so the inference could not have been used to support a prediction of the sex of the child; but it is a perfectly satisfactory causal explanation for all that. The second premiss of the explanatory inference is a causal law. Its converse is also true, and can be used in conjunction with the conclusion of the explanatory inference to deduce the first premiss of that inference. Indeed, that is how I know the genotype of the successful germ cell! But the deduction which tells me that is no part of the explanatory inference; it is rather part of the business of verifying that the explanatory inference really does meet the conditions for being a causal explanation. ('The premisses of demonstrated knowledge must be true' among other things.) The causal and statistical explanations can be made to match by referring the statistics not to the sex of the child, but to the genotype of the relevant sperm cell: the usual process whereby spermatozoa unite with ova is a stochastic one in which the statistical probability is 1/2 that the winner will have the *Y* genotype.

This is a case where before the outcome is known, we know that one of two causal explanations of the outcome will be correct – and we shall know *which* explanation is correct when we know the outcome. Before the event we had a lottery with causal explanations as prizes.

The same sort of thing can happen after the event, as in this made-up example: I draw at random from a box containing a two-headed penny and a normal one, toss, observe that a head turned up, and return the penny to the box without examining it further. Here the lottery is between a statistical and a causal explanation; the outcome of the lottery is not settled by the outcome of the toss, as it would have been if a tail had turned up; and we end in a position where as far as we know, the question 'Why' may or may not have an answer!

Finally, consider the question of why my car would not start this morning. Here the question surely has an answer: there *is* a causal explanation although I do not know what it is. Suppose that it is only one morning in about twenty that my car will not start, and suppose that when it *does not* start, then nine times out of ten it is because of the defective frammas which I never had fixed. Here the statistical component generates an epistemic lottery in which the prizes are causes. There is an answer to the question 'Why?' and the statistics give us a strong clue: it is very likely to be the frammas. Indeed, the probability that it is the frammas is .9, but there is no temptation here to speak of a statistical explanation of strength .9. Rather, the statistical component of the explanation goes like this: non-starting is the product of a stochastic process according to which the frammas goes bad with probability (1/20) (9/10) = .045 and something else goes bad with probability (1/20) (1/10) = .005, and in either case the car definitely is prevented from starting. Given this stochastic law, the probability of non-starting is only .05, and this would be the strength of the Hempelian inferential explanation of the non-starting. But the description of the stochastic process is better than the number .05 would indicate, as a statistical explanation: as far as it goes, it is perfect, and it goes as far as statistics will take us. What is needed to round it out is an item of causal knowledge: was it the frammas?

CONCLUSION

Sometimes statistics enter our understanding of phenomena *via* causal lotteries. In such cases the full explanation would have both a statistical and causal component; and it is doubtful whether Hempel would want to apply his statistical inferential model in such cases. Among the cases where he would apply his model, some are of the 'beautiful' variety where the strength of the inference is so great as to allow us to speak of practical certainty in this sense: the probability of the phenomenon to be explained is so high, given the stochastic law governing the process that produces it, as to make no odds in any gamble or deliberation. These are the cases where the inferential model of explanation seems unexceptionable. But where the strength of the inference is more modest, I think it simply wrong to view the inference as an explanation, and to identify the strength of the inference with the strength of the explanation. To explain the phenomenon that there was at least one head in two tosses of a coin, I would point out that the process is stochastic with probability $\frac{1}{2}$ of head on each toss, and with different tosses independent of one another. I would give the same explanation if matters turned out differently: if, improbably, there had been no head on either toss. The difference between the two cases would lie entirely with the gloss: in the first case one would point out that the probable happened, while in the second one would point out that the improbable happened. But the strength of the explanation would be the same in either case. Finally I point out that since the probable happens more often than not, we are usually able to provide a Hempelian inference when we are able to give a statistical explanation; and I suggest that this is what gives the view that statistical explanations are statistical inferences its specious plausibility.

NOTES

[1] Book 1, chapter 2. All citations from this work are from the Oxford translation.
[2] I leave to one side the deductive-statistical explanations discussed in Part 3.2 of *Aspects of Scientific Explanation*, The Free Press, New York, 1965.
[3] The notion of practical certainty operative in this account of the 'beautiful' cases of statistical explanation is problematical but, I think, defensible. Example: There would be serious trouble with that notion if, given a probability greater than 0 (no matter how slightly greater), one could name a prize so great that the prospect of getting that prize with that probability is not negligible. But I take it that such problems as the St. Petersburg paradox already force us to realize that if Bayesian decision-theory

is to work, there must be a finite upper bound on the utilities of the things the agent can envisage as prizes.

[4] The lines from Louis MacNeice are the last two of his 'Bagpipe Music'.

[5] I would prefer to avoid this second kind of probability and speak simply of the statistical probability of the conclusion; indeed, $p(C)$ is $1-2^{-n}$, i.e., the probability measure defined in the premiss assigns to the conclusion precisely the value which the inductive probability measure c assigns to the conclusion conditionally on the premiss. But let us stay with Hempel's way of talking.

Statistical Explanation

WESLEY C. SALMON

Indiana University

EVER SINCE HIS CLASSIC PAPER with Paul Oppenheim, "Studies in the Logic of Explanation," first published in 1948,[1] Carl G. Hempel has maintained that an "explanatory account [of a particular event] may be regarded as an argument to the effect that the event to be explained . . . *was to be expected* by reason of certain explanatory facts" (my italics).[2] It seems fair to say that this basic principle has guided Hempel's work on *inductive* as well as *deductive* explanation ever since.[3] In spite of its enormous intuitive appeal, I believe that this precept is incorrect and that it has led to an unsound account of scientific explanation. In this paper I shall attempt to develop a different account of explanation and argue for its superiority over the Hempelian one. In the case of inductive explanation, the difference between the two treatments hinges fundamentally upon the question of whether the relation between the explanans and the explanadum is to be understood as a relation of *high*

This paper grew out of a discussion of statistical explanation presented at the meeting of the American Association for the Advancement of Science, held in Cleveland in 1963, as a part of the program of Section L organized by Adolf Grünbaum, then vice-president for Section L. My paper, "The Status of Prior Probabilities in Statistical Explanation," along with Henry E. Kyburg's comments and my rejoinder, were published in *Philosophy of Science,* XXXII, no. 2 (April 1965). The original version of this paper was written in 1964 in an attempt to work out fuller solutions to some problems Kyburg raised, and it was presented at the Pittsburgh Workshop Conference in May 1965, prior to the publication of Carl G. Hempel, *Aspects of Scientific Explanation* (New York: Free Press, 1965).

I should like to express my gratitude to the National Science Foundation for support of the research contained in this paper.

probability or as one of *statistical relevance*. Hempel obviously chooses the former alternative; I shall elaborate an account based upon the latter one. These two alternatives correspond closely to the "concepts of firmness" and the "concepts of increase of firmness," respectively, distinguished by Rudolf Carnap in the context of confirmation theory.[4] Carnap has argued, convincingly in my opinion, that confusion of these two types of concepts has led to serious trouble in inductive logic; I shall maintain that the same thing has happened in the theory of explanation. Attention will be focused chiefly upon inductive explanation, but I shall try to show that a similar difficulty infects deductive explanation and that, in fact, deductive explanation can advantageously be considered as a special limiting case of inductive explanation. It is my hope that, in the end, the present *relevance* account of scientific explanation will be justified, partly by means of abstract "logical" considerations and partly in terms of its ability to deal with problems that have proved quite intractable within the Hempelian schema.

1. The Hempelian Account

Any serious contemporary treatment of scientific explanation must, it seems to me, take Hempel's highly developed view as a point of departure. In the famous 1948 paper, Hempel and Oppenheim offered a systematic account of deductive explanation, but they explicitly denied that all scientific explanations fit that pattern; in particular, they called attention to the fact that some explanations are of the inductive variety. In spite of fairly general recognition of the need for inductive explanations, even on the part of proponents of Hempel's deductive model, surprisingly little attention has been given to the problem of providing a systematic treatment of explanations of this type. Before 1965, when he published "Aspects of Scientific Explanation,"[5] Hempel's "Deductive-Nomological vs. Statistical Explanation"[6] was the only well-known extensive discussion. One could easily form the impression that most theorists regarded deductive and inductive explanation as quite similar in principle, so that an adequate account of inductive explanation would emerge almost routinely by replacing the universal laws of deductive explanation with statistical generalizations, and by replacing the deductive relationship between explanans and explanandum with some sort of inductive relation. Such an attitude was, of course, dangerous in the extreme, for even our present limited knowledge of inductive logic points to deep and fundamental differences between deductive and inductive logical relations. This fact should have made us quite wary of drawing casual analogies between deductive and inductive patterns of explanation.[7] Yet even Hempel's detailed examination of statistical explanation [8] may have contributed to the false feeling of security, for one of the most significant results of that study was that both deductive and inductive explanations must fulfill a *requirement of total evidence*. In the case of deductive explanations the requirement is automatically satisfied; in the case of inductive explanations that requirement is nontrivial.

Accordingly, the situation in May 1965, at the time of the Pittsburgh Workshop Conference, permitted a rather simple and straightforward characterization which would cover both deductive and inductive explanations of particular events.[9] Either type of explanation, according to Hempel, is an argument; as such, it is a linguistic entity consisting of premises and conclusion.[10] The premises constitute the explanans, and the conclusion is the explanandum. The term "explanadum event" may be used to refer to the fact to be explained; the explanandum is the statement asserting that this fact obtains. The term "explanatory facts" may be used to refer to the facts adduced to explain the explanandum event; the explanans is the set of statements asserting that these explanatory facts obtain.[11] In order to explain a particular explanandum event, the explanatory facts must include both particular facts and general uniformities. As Hempel has often said, general uniformities as well as particular facts can be explained, but for now I shall confine attention to the explanation of particular events.

The parallel between the two types of explanation can easily be seen by comparing examples; here are two especially simple ones Hempel has offered: [12]

(1) *Deductive*
This crystal of rock salt, when put into a Bunsen flame, turns the flame yellow, for it is a sodium salt, and all sodium salts impart a yellow color to a Bunsen flame.

(2) *Inductive*
John Jones was almost certain to recover quickly from his streptococcus infection, for he was given penicillin, and almost all cases of streptococcus infection clear up quickly upon administration of penicillin.

These examples exhibit the following basic forms:

(3) *Deductive*

All F are G.
x is F.
———————
x is G.

(4) *Inductive*

Almost all F are G.
x is F.
═══════
x is G.

There are two obvious differences between the deductive and inductive examples. First, the major premise in the deductive case is a universal generalization, whereas the major premise in the inductive case is a statistical generalization. The latter generalization asserts that a high, though unspecified, proportion of F are G. Other statistical generalizations may specify the exact numerical value. Second, the deductive schema represents a valid deductive argument, whereas the inductive schema represents a correct inductive argument. The double line in (4) indicates that the conclusion "follows inductively," that is, with high inductive probability. Hempel has shown forcefully that (4) is *not* to be construed as a deduction with the conclusion that "x is almost certain to be G." [13]

By the time Hempel had provided his detailed comparison of the two types of explanation, certain well-known conditions of adequacy had been spelled out; they would presumably apply both to deductive and to inductive explanations: [14]

(i) The explanatory argument must have correct (deductive or inductive) logical form. In a correct deductive argument the premises entail the conclusion; in a correct inductive argument the premises render the conclusion highly probable.
(ii) The premises of the argument must be true.[15]
(iii) Among the premises there must occur essentially at least one lawlike (universal or statistical) generalization.[16]
(iv) The requirement of total evidence (which is automatically satisfied by deductive explanations that satisfy the condition of validity) must be fulfilled.[17]

Explanations that conform to the foregoing conditions certainly satisfy Hempel's general principle. If the explanation is deductive, the explanandum event was to be expected because the explanandum is deducible from the explanans; the explanans necessitates the explanandum. If the explanation is inductive, it "*explains* a given phenomenon by showing that, in view of certain particular facts and certain statistical laws, its occurrence was to be expected with high logical, or inductive, probability." [18] In this case the explanandum event was to be expected because the explanans confers high probability upon the explanandum; the explanatory facts make the explanandum event highly probable.

2. Some Counterexamples

It is not at all difficult to find cases that satisfy all of the foregoing requirements, but that certainly cannot be regarded as genuine explanations. In a previously mentioned paper [19] I offered the following inductive examples:

(5) John Jones was almost certain to recover from his cold within a week, because he took vitamin C, and almost all colds clear up within a week after administration of vitamin C.

(6) John Jones experienced significant remission of his neurotic symptoms, for he underwent extensive psychoanalytic treatment, and a substantial percentage of those who undergo psychoanalytic treatment experience significant remission of neurotic symptoms.

Both of these examples correspond exactly with Hempel's inductive example (2) above, and both conform to his schema (4). The difficulty with (5) is that colds tend to clear up within a week regardless of the medication administered, and, I understand, controlled tests indicate that the percentage of recoveries is unaffected by the use of vitamin C.[20] The problem with (6) is the substantial spontaneous remission rate for neurotic symptoms of individuals who undergo no psychotherapy of any kind. Before we accept (6) as having any explanatory value whatever, we must know whether the remission rate for psychoanalytic patients is any greater than the spontaneous remission rate. I do not have the answer to this factual question.

I once thought that cases of the foregoing sort were peculiar to inductive explanation, but Henry Kyburg has shown me to be mistaken by providing the following example:

(7) This sample of table salt dissolves in water, for it has had a dissolving spell cast upon it, and all samples of table salt that have had dissolving spells cast upon them dissolve in water.[21]

It is easy to construct additional instances:

(8) John Jones avoided becoming pregnant during the past year, for he has taken his wife's birth control pills regularly, and every man who regularly takes birth control pills avoids pregnancy.

Both of these examples correspond exactly with Hempel's deductive example (1), and both conform to his schema (3) above. The difficulty with (7) and (8) is just like that of the inductive examples (5) and (6). Salt dissolves, spell or no spell, so we do not need to explain the dissolving of this sample in terms of a hex. Men do not become pregnant, pills or no pills, so the consumption of oral contraceptives is not required to explain the phenomenon in John Jones's case (though it may have considerable explanatory force with regard to his wife's pregnancy).

Each of the examples (5) through (8) constitutes an argument to show that the explanandum event was to be expected. Each one has correct (deductive or inductive) logical form—at least it has a form provided by one of Hempel's schemata.[22] Each one has true premises (or so we may assume for the present discussion). Each one has a (universal or statistical) generalization among its premises, and there is no more reason to doubt the lawlikeness of any of these generalizations than there is to doubt the lawlikeness of the generalizations in Hempel's examples. In each case the general premise is essential to the argument, for it would cease to have correct logical form if the general premise were simply deleted. We may assume that the requirement of total evidence is fulfilled in all cases. In the deductive examples it is automatically satisfied, of course, and in the inductive examples we may safely suppose that there is no further available evidence which would alter the probability of John Jones's recovery from either his cold or his neurotic symptoms. There is nothing about any of these examples, so far as I can see, which would disqualify them in terms of the foregoing criteria without also disqualifying Hempel's examples as well.

Folklore, ancient and modern, supplies further instances that would qualify as explanatory under these conditions:

(9) The moon reappeared after a lunar eclipse, for the people made a great deal of noise, banging on pots and pans and setting off fireworks, and the moon always reappears after an eclipse when much noise occurs.[23] (Ancient Chinese folklore)

(10) An acid-freak was standing at the corner of Broadway and Forty-second Street, uttering agonizing moans, when a passerby asked him what was the matter. "Why, man, I'm keeping the wild tigers away," he answered. "But there are no wild tigers

around here," replied the inquirer. "Yeah, see what a good job I'm doing!" (Modern American folklore)

A question might be raised about the significance of examples of the foregoing kind on the ground that conditions of adequacy are normally understood to be necessary conditions, not sufficient conditions. Thus, the fact that there are cases which satisfy all of the conditions of adequacy but which are not explanations is no objection whatever to the necessity of such conditions; this fact could be construed as an objection only if the conditions of adequacy are held to be jointly sufficient. In answer to this point, it must be noted that, although the Hempel-Oppenheim paper of 1948 did begin by setting out conditions of adequacy, it also offered a definition of deductive-nomological explanation, that is, explicit necessary and sufficient conditions.[24] The deductive examples offered above satisfy that definition. And if that definition is marred by purely technical difficulties, it seems clear that the examples conform to the spirit of the original definition.

One might, however, simply acknowledge the inadequacy of the early definition and maintain only that the conditions of adequacy should stand as necessary conditions for satisfactory explanations. In that case the existence of a large class of examples of the foregoing sort would seem to point only to the need for additional conditions of adequacy to rule them out. Even if one never hopes to have a set of necessary conditions that are jointly sufficient, a large and important set of examples which manifest a distinctive characteristic disqualifying them as genuine explanations demands an additional condition of adequacy. Thus, it might be said, the way to deal with the counterexamples is by adding a further condition of adequacy, much as Hempel has done in "Aspects of Scientific Explanation," where he enunciates the *requirement of maximal specificity*.[25] I shall argue, however, that this requirement does not exclude the examples I have offered, although a minor emendation will enable it to do so. Much more significantly, I shall maintain that the requirement of maximal specificity is the wrong type of requirement to take care of the difficulties that arise and that a much more fundamental revision of Hempel's conception is demanded. However, before beginning the detailed discussion of the requirement of maximal specificity, I shall make some general remarks about the counterexamples and their import—especially the inductive ones—in order to set the stage for a more enlightening discussion of Hempel's recent addition to the set of conditions of adequacy.

3. Preliminary Analysis

The obvious trouble with our horrible examples is that the "explanatory" argument is not needed to make us see that the explanandum event was to be expected. There are other, more satisfactory, grounds for this expectation. The "explanatory facts" adduced are irrelevant to the explanandum event despite the fact that the explanandum follows (deductively or inductively) from the explanans. Table salt dissolves in water regardless of hexing, almost all colds clear up within a week regardless of treatment, males do not get pregnant regardless of pills, the moon reappears regardless of the amount of Chinese din, and there are no wild tigers in Times Square regardless of our friend's moans. Each of these explanandum events has a high prior probability independent of the explanatory facts, and the probability of the explanandum event relative to the explanatory facts is the same as this prior probability. In this sense the explanatory facts are irrelevant to the explanandum event. The explanatory facts do nothing to enhance the probability of the explanandum event or to make us more certain of its occurrence than we would otherwise have been. This is not because we know that the fact to be explained has occurred; it is because we had other grounds for expecting it to occur, *even if we had not already witnessed it.*

Our examples thus show that it is not correct, even in a preliminary and inexact way, to characterize explanatory accounts as arguments showing that the explanandum event was to be expected. It is more accurate to say that an explanatory argument shows that the probability of the explanandum event relative to the explanatory facts is substantially greater than its prior probability.[26] An explanatory account, on this view, increases the degree to which the explanandum event was to be expected. As will emerge later in this paper, I do not regard such a statement as fully accurate; in fact, the increase in probability is merely a pleasant by-product which often accompanies a much more fundamental characteristic. Nevertheless, it makes a useful starting point for further analysis.

We cannot, of course, hope to understand even the foregoing rough characterization without becoming much clearer on the central concepts it employs. In particular, we must consider explicitly what is meant by "probability" and "prior probability." There are several standard views on the nature of probability, and I propose to consider briefly how the foregoing characterization of explanation, especially inductive explanation, would look in terms of each.

a. According to the *logical interpretation*, probability or degree of confirmation is a logical relation between evidence and hypothesis.[27] Degree of confirmation statements are analytic. There is no analogue of the rule of detachment (*modus ponens*), so inductive logic does not provide any basis for asserting inductive conclusions. There are certain methodological rules for the application of inductive logic, but these are not part of inductive logic itself. One such rule is the requirement of total evidence. For any given explanandum, taken as hypothesis, there are many true degree of confirmation statements corresponding to different possible evidence statements. The probability value to be chosen for practical applications is that given by the degree of confirmation statement embodying the total available (relevant) evidence. This is the value to be taken as a fair betting quotient, as an estimate of the relative frequency, or as the value to be employed in calculating the estimated utility.

I have long been seriously puzzled as to how Carnap's conception of logical probability could be consistently combined with Hempel's views on the nature of explanation.[28] In the first place, for purposes of either deductive or inductive explanation, it is not clear how the universal or statistical generalizations demanded by Hempel's schemata are to become available, for inductive logic has no means of providing such generalizations as detachable conclusions of any argument. In the second place, degree of confirmation statements embodying particular hypothesis statements and particular evidence statements are generally available, so inductive "inferences" from particulars to particulars are quite possible. In view of this fact, it is hard to see why inductive explanation should require any sort of general premise. In the third place, the relation between explanans and explanandum would seem to be a degree of confirmation statement, and not an argument with premises and conclusions in the usual sense. This last point seems to me to be profoundly important. Carnap's recognition that much, if not all, of inductive logic can proceed without inductive arguments—that is, without establishing conclusions on the basis of premises—is a deep insight that must be reckoned with. As a consequence, we must seriously question whether an explanation is an argument at all. This possibility arises, as I shall try to show later, even if one does not adhere to a logical interpretation of probability. Indeed, within an account of explanation in terms of the frequency interpretation, I shall argue contra Hempel that *explanations are not arguments.*[29]

For the logical interpretation of probability, it would seem, an explanation should involve an addition of new evidence to the body of total evidence resulting in a new body of total evidence relative to which the probability of the explanandum is higher than it was relative to the old body of total evidence. In the usual situation, of course, an explanation is sought only after it is known that the explanandum event has occurred. In this case, the body of total evidence already contains the explanandum, so no addition to the body of total evidence can change its probability. We must, therefore, somehow circumscribe a body of total evidence available prior to the occurrence of the explanandum event, relative to which our prior probability is to be taken. This body of evidence must contain neither the explanans nor the explanandum.[30]

The logical interpretation of probability admits degrees of confirmation on tautological evidence, and these probabilities are not only prior but also a priori. But a priori prior probabilities are not especially germane to the present discussion. We are not concerned with the a priori probability of a Bunsen flame turning yellow, nor with the a priori probability of a cold clearing up in a week, nor with the a priori probability of a remission of ones neurotic symptoms, nor with the a priori probability of finding a wild tiger in Times Square, etc. We are concerned with the probabilities of these events relative to more or less specific factual evidence. The prior probabilities are not logically a priori; they are prior with respect to some particular information or investigation.

b. According to the new *subjective* or *personalistic interpretation*, probability is simply orderly opinion.[31] The required orderliness is provided by the mathematical calculus of probability. One of the main concerns of the personalists is with the revision of opinion or degree of belief in the light of new evidence. The account of explanation suggested above fits very easily with the personalistic interpretation. At some time before the explanandum event occurs, the personalist would say, an individual has a certain degree of belief that it will occur—reflected, perhaps, in the kind of odds he is willing to give or take in betting on its occurrence. According to the present view of explanation, a personalist should require that the explanatory facts be such as to increase the prior confidence in the occurrence of the explanandum event. Of course, this view deliberately introduces into its treatment of scientific explanation a large degree of subjectivity, but this is not necessarily a defect in either the account of probability or the account of explanation.

c. According to the *frequency interpretation*, a probability is the limit of the relative frequency of an attribute in an infinite sequence of events.[32] A probability cannot, therefore, literally be assigned to a single event. Since there are many occasions when we must associate probabilities with single events, and not only with large (literally, infinite) aggregates of events, a way must be found for making the probability concept applicable to single events. We wager on single events—a single toss of the dice, a single horse race, a single deal of the cards. The results of our practical planning and effort depend upon the outcomes of single occurrences, not infinite sequences of them. If probability is to be a guide of life, it must be meaningful to apply probability values to single events. This problem of the single case has traditionally been one of the difficulties confronting the frequency theory, and we shall have to examine in some detail the way in which it can be handled. With regard to the single case, the frequency interpretation is on an entirely different footing from the logical and subjective interpretations. Neither of these latter interpretations faces any special problem of the single case, for statements about single events are perfectly admissible hypotheses for the logical theory, and the subjective theory deals directly with degrees of belief regarding single events.

The central topic of concern in this paper is the explanation of single events. If the frequency theory is to approach this problem at all, it must deal directly with the problem of the single case. To some, the fact that the frequency interpretation is faced with this special problem of the single case may constitute a compelling reason for concluding that the frequency interpretation is totally unsuitable for handling the explanation of single events. Such a conclusion would, I think, be premature. On the contrary, I would argue that a careful examination of the way in which the frequency interpretation handles the single case should prove extremely illuminating with respect to the general problem of inductive explanation. Although I have strong prejudices in favor of the frequency interpretation, this paper is, nevertheless, not the place to argue for them.[33] In this context I would claim instead that frequencies play such an important role in any theory of probability that an examination of the problem of the single case cannot fail to cast light on the problem of explanation regardless of one's persuasion concerning interpretations of the probability concept.

d. In recent years there has been a good deal of discussion of the *propensity interpretation* of probability.[34] This interpretation is so similar to the frequency interpretation in fundamental respects that everything I shall say about the problem of the single case for the frequency interpretation can be made directly applicable to the propensity interpretation by a simple translation: wherever I speak of "the problem of selecting the appropriate reference class" in connection with the frequency interpretation, read "the problem of specifying the nature of the chance setup" in reference to the propensity interpretation. It seems to me that precisely parallel considerations apply for these two interpretations of probability.

4. The Single Case

Let *A* be an unending sequence of draws of balls from an urn, and let *B* be the class of red things. *A* is known as the *reference class,* and *B* the *attribute class.* The probability of red draws from this urn, $P(A,B)$, is the limit of the relative frequency with which members of the reference class belong to the attribute class, that is, the limit of the relative frequency with which draws from the urn result in a red ball as the number of draws increases without any bound.[35]

Frequentists like John Venn and Hans Reichenbach have dealt with the problem of the single case by assigning each single event to a reference class and by transferring the probability value from that reference class to the single event in question.[36] Thus, if the limit of the relative frequency of red among draws from our urn is one-third, then we say that the probability of getting red on *the next draw* is one-third. In this way the meaning of the probability concept has been extended so that it applies to single events as well as to large aggregates.

The fundamental difficulty arises because a given event can be referred to any of a large number of reference classes, and the probability of the attribute in question may vary considerably from one of these to another. For instance, we could place two urns on a table, the one on the left containing only red balls, the one on the right containing equal numbers of red, white, and blue balls. The reference class *A* might consist of blind drawings from the right-hand urn, the ball being replaced and the urn thoroughly shaken after each draw. Another reference class *A′* might consist of draws made alternately from the left- and the right-hand urns. Infinitely many other reference classes are easily devised to which the next draw—the draw with which we are concerned—belongs. From which reference class shall we transfer our probability value to this single case? A method must be established for choosing the appropriate reference class. Notice, however, that there is no difficulty in selecting an attribute class. The question we ask determines the attribute class. We want to know the probability of getting red, so there is no further problem about the attribute class.

Reichenbach recommends adopting as a reference class "the narrowest class for which reliable statistics can be compiled."[37] This principle is, as Reichenbach himself has observed, rather ambiguous. Since increasing the reliability of statistics generally tends to broaden the class and since narrowing the class often tends to reduce the reliability of the statistics, the principle involves two desiderata which pull in opposite directions. It seems that we are being directed to maximize two variables that cannot simultaneously be maximized. This attempt to extend the meaning of the probability concept to single cases fails to provide a method for associating a unique probability value with a given single event. Fully aware of this fact, Reichenbach insisted that the probability concept applies *literally* only to sequences; talk about the probability of a single event is "elliptical" and the extended meaning is "fictitious." The choice of a reference class, he maintained, is often dictated by practical rather than theoretical considerations.

Although Reichenbach has not said so, it seems reasonable to suppose that he was making a distinction similar to that made by Carnap between the principles belonging to inductive logic and methodological rules for the application of inductive logic.[38] The requirement of total evidence, it will be recalled, is a methodological rule for the application of inductive logic. Reichenbach could be interpreted as suggesting analogously that probability theory itself is concerned only with limit statements about relative frequencies in infinite sequences of events, whereas the principle for selection of a reference class stands as a methodological rule for the practical application of probability statements. (A much stronger analogy between the requirement of total evidence and the principle for choosing a reference class for a single case will be shown below.) In fact, for Reichenbach, it might have been wise to withhold the term "probability" from single events, reserving his term "weight" for this purpose. We could then say that practical considerations determine what probability should be chosen to serve as a weight for a particular single event. The relative practical importance of reliability and precision would then determine the extent to which narrowness gives way to reliability of statistics (or conversely) in determining the appropriate reference class.

Although Reichenbach's formulation of the principle for the selection of reference classes is not entirely satisfactory, his intention seems fairly clear. In order to transfer a probability value from a sequence to a single case, it is necessary to have some basis for ascertaining the probability in that sequence. The reference class must, therefore, be broad enough to provide the required number of instances for examination to constitute evidence for an inductive inference. At the same time, we want to avoid choosing a reference class so broad that it includes cases irrelevant to the one with which we are concerned.

Statistical relevance is the essential notion here. It is desirable to narrow the reference class in statistically relevant ways, but not in statistically irrelevant ways. When we choose a reference class to which to refer a given single case, we must ask whether there is any statistically relevant way to subdivide that class. If so, we may choose the narrower subclass that results from the subdivision; if no statistically relevant way is known, we must avoid making the reference class any narrower. Consider, for example, the probability that a particular individual, John Smith, will still be alive ten years hence. To determine this probability, we take account of his age, sex, occupation, and health; we ignore his eye color, his automobile license number, and his last initial. We expect the relative frequency of survival for ten more years to vary among the following reference classes: humans, Americans, American males, forty-two-year-old American males, forty-two-year-old American male steeplejacks, and forty-two-year-old American male steeplejacks suffering from advanced cases of lung cancer. We believe that the relative frequency of survival for another ten years is the same in the following classes: forty-two-year-old American male steeplejacks with advanced cases of lung cancer, forty-two-year-old blue-eyed American male steeplejacks with advanced cases of lung cancer, and forty-two-year-old blue-eyed American male steeplejacks with even automobile license plate numbers who suffer from advanced cases of lung cancer.

Suppose we are dealing with some particular object or event x, and we seek to determine the probability (weight) that it has attribute B. Let x be assigned to a reference class A, of which it is a member. $P(A,B)$ is the probability of this attribute within this reference class. A set of mutually exclusive and exhaustive subclasses of a class is a *partition* of that class. We shall often be concerned with partitions of reference classes into two subclasses; such partitions can be effected by a property C which divides the class A into two subclasses, $A.C$ and $A.\overline{C}$. A property C is said to be *statistically relevant* to B within A if and only if $P(A.C,B) \neq P(A,B)$. This notion of statistical relevance is the fundamental concept upon which I hope to build an explication of inductive explanation.

In his development of a frequency theory based essentially upon the concept of randomness, Richard von Mises introduced the notion of a *place selection:* "By a place selection we mean the selection of a partial sequence in such a way that we decide whether an element should or should not be included without making use of the attribute of the element." [39] A place selection effects a partition of a reference class into two subclasses, elements of the place selection and elements not included in the place selection. In the reference class of draws from our urn, every third draw starting with the second, every *k*th draw where *k* is prime, every draw following a red result, every draw made with the left hand, and every draw made while the sky is cloudy all would be place selections. "Every draw of a red ball" and "every draw of a ball whose color is at the opposite end of the spectrum from violet" do not define place selections, for membership in these classes cannot be determined without reference to the attribute in question.

A place selection may or may not be statistically relevant to a given attribute in a given reference class. If the place selection is statistically irrelevant to an attribute within a reference class, the probability of that attribute within the subclass determined by the place selection is equal to the probability of that attribute within the entire original reference class. If every place selection is irrelevant to a given attribute in a given sequence, von Mises called the sequence *random.* If every property that determines a place selection is statistically irrelevant to *B* in *A*, I shall say that *A* is a *homogeneous reference class* for *B*. A reference class is homogeneous if there is no way, even in principle, to effect a statistically relevant partition without already knowing which elements have the attribute in question and which do not. Roughly speaking, each member of a homogeneous reference class is a random member.

The aim in selecting a reference class to which to assign a single case is not to select the narrowest, but the widest, available class. However, the reference class should be homogeneous, and achieving homogeneity requires making the reference class narrower if it was not already homogeneous. I would reformulate Reichenbach's method of selection of a reference class as follows: choose the broadest homogeneous reference class to which the single event belongs. I shall call this the *reference class rule.*

Let me make it clear immediately that, although I regard the above formulation as an improvement over Reichenbach's, I do not suppose that it removes all ambiguities about the selection of reference classes either in principle or in practice. In principle it is possible for an event to belong to two equally wide homogeneous reference classes, and the probabilities of the attribute in these two classes need not be the same. For instance, suppose that the drawing from the urn is not random and that the limit of the relative frequency of red for every *k*th draw (*k* prime) is 1/4, whereas the limit of the relative frequency of red for every even draw is 3/4. Each of these subsequences may be perfectly random; each of the foregoing place selections may, therefore, determine a homogeneous reference class. Since the intersection of these two place selections is finite, it does not determine a reference class for a probability. The second draw, however, belongs to both place selections; in this fictitious case there is a genuine ambiguity concerning the probability to be taken as the weight of red on the second draw.

In practice we often lack full knowledge of the properties relevant to a given attribute, so we do not know whether our reference class is homogeneous or not. Sometimes we have strong reason to believe that our reference class is not homogeneous, but we do not know what property will effect a statistically relevant partition. For instance, we may believe that there are causal factors that determine which streptococcus infections will respond to penicillin and which ones will not, but we may not yet know what these causal factors are. When we know or suspect that a reference class is not homogeneous, but we do not know how to make any statistically relevant partition, we may say that the reference class is *epistemically homogeneous*. In other cases, we know that a reference class is inhomogeneous and we know what attributes would effect a statistically relevant partition, but it is too much trouble to find out which elements belong to each subclass of the partition. For instance, we believe that a sufficiently detailed knowledge of the initial conditions under which a coin is tossed would enable us to predict (perfectly or very reliably) whether the outcome will be heads or tails, but practically speaking we are in no position to determine these initial conditions or make the elaborate calculations required to predict the outcome. In such cases we may say that the reference class is *practically homogeneous*.[40]

The reference class rule remains, then, a methodological rule for the application of probability knowledge to single events. In practice we attempt to refer our single cases to classes that are practically or epistemically homogeneous. When something important is at stake, we may try to extend our knowledge in order to improve the degree of homogeneity we can achieve. Strictly speaking, we cannot meaningfully refer to degrees of homogeneity until a quantitative concept of homogeneity has been provided. That will be done in section 6 below.

It would, of course, be a serious methodological error to assign a single case to an inhomogeneous reference class if neither epistemic nor practical considerations prevent partitioning to achieve homogeneity. This fact constitutes another basis for regarding the reference class rule as the counterpart of the requirement of total evidence. The requirement of total evidence demands that we use all available relevant evidence; the reference class rule demands that we partition whenever we have available a statistically relevant place selection by means of which to effect the partition.

Although we require homogeneity, we must also prohibit partitioning of the reference class by means of statistically irrelevant place selections. The reason is obvious. Irrelevant partitioning reduces, for no good reason, the inductive evidence available for ascertaining the limiting frequency of our attribute in a reference class that is as homogeneous as we can make it. Another important reason for prohibiting irrelevant partitions will emerge below when we discuss the importance of multiple homogeneous reference classes.

A couple of fairly obvious facts about homogeneous reference classes should be noted at this point. If all *A*'s are *B*, *A* is a homogeneous reference class for *B*. (Somewhat counterintuitively, perhaps, *B* occurs perfectly randomly in *A*.) In this case, $P(A,B) = 1$ and $P(A.C,B) = 1$ for any *C* whatever; consequently, no place selection can yield a probability for *B* different from that in the reference class *A*. Analogously, *A* is homogeneous for *B* if no *A*'s are *B*. In the frequency interpretation, of course, $P(A,B)$ can equal one even though not all *A*'s are *B*. It follows that a probability of one does not entail that the reference class is homogeneous.

Some people maintain, often on a priori grounds, that *A* is homogeneous (not merely practically or epistemically homogeneous) for *B* only if all *A*'s are *B* or no *A*'s are *B;* such people are determinists. They hold that causal factors always determine which *A*'s are *B* and which *A*'s are not *B;* these causal factors can, in principle, be discovered and used to construct a place selection for making a statistically relevant partition of *A*. I do not believe in this particular form of determinism. It seems to me that there are cases in which *A* is a homogeneous reference class for *B* even though not all *A*'s are *B*. In a sample of radioactive material a certain percentage of atoms disintegrate in a given length of time; no place selection can give us a partition of the atoms for which the frequency of disintegration differs from that in the whole sample. A beam of electrons is shot at a potential barrier and some pass through while others are reflected; no place selection will enable us to make a statistically relevant partition in the class of electrons in the beam. A beam of silver atoms is sent through a strongly inhomogeneous magnetic field (Stern-Gerlach experiment); some atoms are deflected upward and some are deflected downward, but there is no way of partitioning the beam in a statistically relevant manner. Some theorists maintain, of course, that further investigation will yield information that will enable us to make statistically relevant partitions in these cases, but this is, at present, no more than a declaration of faith in determinism. Whatever the final resolution of this controversy, the homogeneity of *A* for *B* does not logically entail that all *A*'s are *B*. The truth or falsity of determinism cannot be settled a priori.

The purpose of the foregoing excursus on the frequency treatment of the problem of the single case has been to set the stage for a discussion of the explanation of particular events. Let us reconsider some of our examples in the light of this theory. The relative frequency with which we encounter instances of water-soluble substances in the normal course of things is noticeably less than one; therefore, the probability of water solubility in the reference class of samples of unspecified substances is significantly less than one. If we ask why a particular sample of unspecified material has dissolved in water, the prior weight of this explanandum event is less than one as referred to the class of samples of unspecified substances. This broad reference class is obviously inhomogeneous with respect to water solubility. If we partition it into the subclass of samples of table salt and samples of substances other than table salt, it turns out that every member of the former subclass is water-soluble. The reference class of samples of table salt is homogeneous with respect to water solubility. The weight for the single case, referred to this homogeneous reference class, is much greater than its prior weight. By referring the explanandum event to a homogeneous reference class and substantially increasing its weight, we have provided an inductive explanation of its occurrence. As the discussion develops, we shall see that the homogeneity of the reference class is the key to the explanation. The increase in weight is a fortunate dividend in many cases.

If we begin with the reference class of samples of table salt, asking why this sample of table salt dissolves in water, we already have a homogeneous reference class. If, however, we subdivide that reference class into hexed and unhexed samples, we have added nothing to the explanation of dissolving, for no new probability value results and we have not made the already homogeneous reference class any more homogeneous. Indeed, we have made matters worse by introducing a statistically irrelevant partition.

The original reference class of samples of unspecified substances can be partitioned into hexed and unhexed samples. If this partition is accomplished by means of a place selection—that is, if the hexing is done without reference to previous knowledge about solubility—the probabilities of water solubility in the subclasses will be no different from the probability in the original reference class. The reference class of hexed samples of unspecified substances is no more homogeneous than the reference class of samples of unspecified substances; moreover, it is narrower. The casting of a dissolving spell is statistically irrelevant to water solubility, so it cannot contribute to the homogeneity of the reference class, and it must not be used in assigning a weight to the single case. For this reason it contributes nothing to the explanation of the fact that this substance dissolves in water.

The vitamin C example involves the same sort of consideration. In the class of colds in general, there is a rather high frequency of recovery within a week. In the narrower reference class of colds for which the victim has taken vitamin C, the frequency of recovery within a week is no different. Vitamin C is not efficacious, and that fact is reflected in the statistical irrelevance of administration of vitamin C to recovery from a cold within a week. Subdivision of the reference class in terms of administration of vitamin C does not yield a more homogeneous reference class and, consequently, does not yield a higher weight for the explanandum event. In similar fashion, we know that noisemaking and shooting off fireworks are statistically irrelevant to the reappearance of the moon after an eclipse, and we know that our friend's loud moaning is statistically irrelevant to finding a wild tiger in Times Square. In none of these horrible examples do the "explanatory facts" contribute anything toward achieving a homogeneous reference class or to an increase of posterior weight over prior weight.

5. The Requirement of Maximal Specificity

In his 1962 essay, "Deductive-Nomological vs. Statistical Explanation," Hempel takes note of what he calls "the ambiguity of statistical systematization,"[41] and he invokes a requirement of total evidence to overcome it.[42] This kind of ambiguity is seen to arise out of the possibility of referring a given event to several different reference classes, and Hempel explicitly compares his requirement of total evidence to Reichenbach's rule of selecting the narrowest reference class for which reliable statistics are available.[43] These requirements do not exclude any of the counterexamples introduced above, in particular, the "explanations" of John Jones's recovery from his cold (5) and of his remission of neurotic symptoms (6).

In the 1965 essay, "Aspects of Scientific Explanation," Hempel distinguishes the ambiguity of statistical explanation from the epistemic ambiguity of statistical explanation. The former ambiguity arises from the existence of different reference classes, with different relative frequencies of the attribute in question, to which a given single case may be referred. The latter ambiguity arises from the occurrence of items in our body of accepted scientific knowledge which lead us to refer the given single case to different reference classes. In order to cope with the problem of epistemic ambiguity, Hempel introduces the *requirement of maximal specificity for inductive-statistical explanations* as follows:

Consider a proposed explanation of the basic statistical form

$$\text{(30)} \qquad \frac{\begin{array}{c} p(G,F) = r \\ Fb \end{array}}{Gb} \quad r]$$

Let s be the conjunction of the premises, and, if K is the set of all statements accepted at the given time, let k be a sentence that is logically equivalent to K (in the sense that k is implied by K and in turn implies every sentence in K). Then, to be rationally acceptable in the knowledge situation represented by K, the proposed explanation (30) must meet the following condition (the requirement of maximal specificity): If $s.k$ implies that b belongs to a class F_1, and that F_1 is a subclass of F, then $s.k$ must also imply a statement specifying the statistical probability of G in F_1, say

$$p(G,F_1) = r_1$$

Here, r_1 must equal r unless the probability statement just cited is simply a theorem of mathematical probability theory.[44]

Hempel goes on to remark that

the requirement of maximal specificity, then, is here tentatively put forward as characterizing the extent to which the requirement of total evidence properly applies to inductive-statistical explanations. The general idea thus suggested comes to this: In formulating or appraising I-S explanation, we should take into account all that information provided by K which is of potential *explanatory* relevance to the explanandum event; i.e., all pertinent statistical laws, and such particular facts as might be connected, by the statistical laws, with the explanandum event.[45]

There are two immediate objections to the requirement of maximal specificity as thus formulated. First, like the requirement of total evidence, it fails to rule out such counterexamples as (5), the explanation of John Jones's recovery from his cold on account of taking vitamin C. This "explanation" is not blocked by the requirement, for, as we have noted, there is no narrower class for which the probability of recovery from a cold would be different. The trouble is that the class invoked in the "explanation" is too narrow; the class of people with colds—regardless of medication—already satisfies the requirement. The difficulty seems to arise from the emphasis upon *narrowness* in Reichenbach's formulation of his reference class rule. As I reformulate the rule, we seek the broadest homogeneous reference class. However, this problem is not serious, for it can be circumvented by a minor reformulation of Hempel's requirement. We need simply to add to Hempel's formulation the condition that F is the largest class that satisfies it. This addition would result in what might be called the *requirement of the maximal class of maximal specificity.*

Second, the final sentence in the requirement seems designed to achieve epistemic homogeneity, but it is not strong enough to do so. In explaining the reason for the qualification "r_1 must equal r unless the probability statement just cited is simply a theorem of mathematical probability theory," Hempel points out that $F.G$ is always a subset of F, but obviously $p(G,F.G) = 1$ by the mathematical calculus alone. Hence, without the qualification of the "unless" clause, the requirement of maximal specificity could never be nontrivially fulfilled.

When von Mises attempted to characterize randomness in precisely the sense that is at issue here, he pointed out that the relative frequency in a subsequence must be invariant if the subsequence is determined by a place selection, but that it need not be invariant for subsequences determined by other types of selections. For instance, let F be the class of draws from a particular urn and G the drawing of a red ball. Suppose, as a matter of fact, that the person drawing from the urn sometimes draws with his left hand and sometimes with his right, and that every draw with the left hand produces a red ball, whereas some draws with the right produce balls of other colors. Let H be the class of draws with the left hand. Under these conditions F does not fulfill the requirement of maximal specificity, for $F.H$ is a subclass of F in which G has a different probability than it does in F. Obviously, the fact that all H's are G is not a theorem of the probability calculus; it is a physical fact.

Now, to take a different example for purposes of comparison, suppose that everything is the same as in the foregoing case, except that $p(G,F) = p(G,H.F)$—that is, the frequency with which red appears is the same regardless of which hand is used for the draw, and, so far as H

is concerned, F satisfies the requirement of maximal specificity. Let J be the class of draws of a ball whose color is at the opposite end of the visible spectrum from violet. Clearly, all J's are G, but I consider this to be an extremely highly confirmed matter of fact, not a theorem of the probability calculus or of pure logic. It follows immediately that $p(G,F) \neq p(G,J.F) = 1$. Hence, by Hempel's formulation F still violates the requirement of maximal specificity, and so obviously will any class F except when either all F's are G or no F's are G. In the first case we wanted to say that the class F violates maximal specificity because of the empirical generalization that all $F.H$'s are G; in the second case we were forced unwillingly to admit that F violates maximal specificity because of the empirical generalization that all $F.J$'s are G. The difference is, roughly speaking, that we can know whether a draw is being made with the left hand before we see the result, but we cannot know that a draw results in a color at the opposite end of the visible spectrum from violet until we know what the result is. To characterize this difference precisely might be difficult, but it is required for von Mises's concept of randomness, for my concept of homogeneity, and, I believe, for a satisfactory reformulation of Hempel's requirement of maximal specificity.

When the foregoing two difficulties are overcome, I believe that Hempel's requirement of maximal specificity will become equivalent to my concept of epistemic homogeneity, and, I believe, it will rule out counterexamples of the sort that I have presented above. It will rule out the deductive examples as well as the inductive examples if we construe the deductive examples as limiting cases of inductive explanation and demand that they also satisfy the requirement of the maximal class of maximal specificity. If, however, these two rather straightforward revisions of Hempel's requirement of maximal specificity were all that is needed to fix up all of the difficulties in Hempel's account of explanation of particular events, it would have been an outrageous imposition upon the good humor of the reader to have taken so many pages to say so. The fact is quite otherwise. These rather lengthy discussions have been intended to lay the foundations for further considerations that will lead to far-reaching differences with Hempel's basic conceptions. The point that we have arrived at so far is that requirements (i) through (iv) above (see page 176) still stand, except that (iv) now contains the corrected requirement of the maximal class of maximal specificity. Requirement (i) still stands, and it demands for inductive explanation that the explanandum be made highly probable (to a degree $\geq$ some chosen number r). According to this conception, an explanation is still regarded as an argument to the effect that a certain event was to be expected by virtue of certain explanatory facts. I shall maintain that we have still failed to come to grips adequately with the problem of relevance of the explanans to the explanandum.

6. Prior Weights and Degree of Inhomogeneity

One of the fundamental features of explanation according to the approach I have been suggesting is that it concerns the relations between the prior weight and the posterior weight of the explanandum event. I have attempted to indicate how a frequency theorist may get weights from probabilities, by application of the reference class rule, but so far nothing has been said about the prior probability from which the prior weight is to be taken. Let us now consider that question.

If we look at Hempel's schemata (3) and (4) for deductive and inductive explanation, respectively, we see that the form of the conclusion in each case is "x is G." However, the explanatory question to which the proffered explanation attempts to supply an answer has a more complex form. We do not ask, "Why is this thing yellow?" We ask, "Why is this Bunsen flame yellow?" We do not ask, "Why does this thing disappear?" We ask "Why does this streptococcus infection disappear?" In every case, I think, the question takes the form, "Why is this x which is A also B?" The answer then takes the form, "Because this x is also C." C must be an attribute that is statistically relevant to B within the reference class A.

The explanatory question, it seems to me, furnishes an original reference class A to which the explanandum event is referred. The probability of the attribute B within that reference class A is the *prior weight* for purposes of that explanation; normally, the class A would be epistemically homogeneous prior to the furnishing of the explanation. In cases of genuine explanation, the reference class A is not actually homogeneous with respect to the attribute B, and so a further property C is invoked to achieve a homogeneous (or more nearly homogeneous) reference class $A.C$ for purposes of establishing a *posterior weight*. If the original reference class A is homogeneous, the introduction of an irrelevant characteristic C to narrow the reference class is not only without explanatory value but is actually methodologically detrimental. In a later section of this paper, I shall offer a more formal characterization of such explanations, showing how the prior weight and posterior weight figure in the schema.

We can now introduce a quantitative measure of the degree of inhomogeneity of the reference class A which will, in a sense, indicate what has been achieved when that class is partitioned into homogeneous subclasses. Let the sequence $x_1, x_2, \ldots, x_i, \ldots$ be the ordered set A whose degree of inhomogeneity with respect to the attribute B is to be measured. Assume that there exists a unique homogeneous partition $C_1, \ldots, C_k$ of A with respect to B.[46] Let $P(A,B) = p$, and let $P(A.C_j,B) = p_j$. Assign to each element x_i of A the weight $w_i = p_j$ associated with that compartment of the partition C_j to which x_i belongs, that is, if $x_i \in A.C_j$, then $w_i = p_j = P(A.C_j,B)$. Evidently, $p = w$ is the prior weight of each x_i in A with respect to the attribute B, whereas w_i is its posterior weight in the homogeneous partition of A.

If we say, in effect, that w_i is the correct weight to assign to x_i, then we may say that $p - w_i$ is the "error" that would result from the use of the prior weight instead of the posterior weight. Squaring to make this quantity nonnegative, we may take the squared error $(p - w_i)^2$ as a measure of error. Then

$$\sum_{i=1}^{n} (p - w_i)^2$$

is the cumulative squared error involved in using the prior weight on the first n elements of A, and

$$\frac{1}{n}\sum_{i=1}^{n} (p - w_i)^2$$

is the *mean squared error.* The limit of the mean squared error as n increases without bound will be taken to represent the inhomogeneity of the reference class A with respect to the attribute B, but for reasons to be mentioned in a moment, I shall multiply this number by a constant for purposes of defining degree of inhomogeneity.

If we think of the weights w_i simply as numbers that exhibit a certain frequency distribution, then it is clear that p represents their mean value, that is,

$$\lim_{n\to\infty} \frac{1}{n}\sum_{i=1}^{n} w_i = p.$$

Then

$$\lim_{n\to\infty} \frac{1}{n}\sum_{i=1}^{n} (p - w_i)^2 \doteq \sigma^2$$

is a measure of the dispersion of the w_i known as the *variance* (which is the square of the standard deviation σ).[47]

If the reference class A is already homogeneous, the error is identically zero for every i, and the variance is zero. The maximum degree of inhomogeneity seems to be represented by the case in which B occurs with probability 1/2 in A, but there is a partition of A such that every element of C_1 is B and no element of C_2 is B. These conditions would obtain, for instance, if a sequence of coin tosses consisted of tosses of a two-headed coin randomly interspersed with tosses of a two-tailed coin, each of the two being used with equal frequency. In this case $p = 1/2$, and each w_i is either 0 or 1. Thus, the error on each toss is 1/2, and the mean squared error (for each n) as well as the variance is 1/4. In order to have a convenient measure of degree of homogeneity, let us multiply the variance by 4, so that degree of inhomogeneity ranges from 0 to 1. We can then adopt the following formal definition:

> The degree of inhomogeneity of class A with respect to attribute $B = 4\sigma^2$.

It is worth noting that the transition from an inhomogeneous reference class A to a set of homogeneous subclasses $A.C_j$ provides an increase of information. In the extreme example given above, when we have only the original inhomogeneous reference class of coin tosses, we know only that the probability of a head is 1/2, and we know nothing about whether an individual toss will result in a head or a tail. When the class is partitioned into two subclasses, one consisting of tosses with the two-headed coin and the other consisting of tosses with the two-tailed coin, we have complete information regarding the outcome of each toss; each result can be predicted with certainty.[48]

7. Causal and Statistical Relevance

The attempt to explicate explanation in terms of probability, statistical relevance, and homogeneity is almost certain to give rise to a standard objection. Consider the barometer example introduced by Michael Scriven in a discussion of the thesis of symmetry between explanation and prediction [49]—a thesis whose discussion I shall postpone until a later section. If the barometer in my house shows a sudden drop, a storm may be predicted with high reliability. But the barometric reading is only an indicator; it does not cause the storm and, according to Scriven, it does not explain its occurrence. The storm is caused by certain widespread atmospheric conditions, and the behavior of the barometer is merely symptomatic of them. "In explanation we are looking for a *cause,* an event that not only occurred earlier but stands in a *special relation* to the other event. Roughly speaking, the prediction requires only a correlation, the explanation more." [50]

The objection takes the following form. There is a correlation between the behavior of the barometer and the occurrence of storms. If we take the general reference class of days in the vicinity of my house and ask for the probability of a storm, we get a rather low prior probability. If we partition that reference class into two subclasses, namely, days on which there is a sudden drop in the barometer and days on which there is not, we have a posterior probability of a storm in the former class much higher than the prior probability. The new reference class is far more homogeneous than the old one. Thus, according to the view I am suggesting, the drop in barometric reading would seem to explain the storm.

I am willing to admit that symptomatic explanations seem to have genuine explanatory value in the absence of knowledge of causal relations, that is, as long as we do not know that we are dealing only with symptoms. Causal explanations supersede symptomatic ones when they can be given, and when we suspect we are dealing with symptoms, we look hard for a causal explanation. The reason is that a causal explanation provides a more homogeneous reference class than does a symptomatic explanation. Causal proximity increases homogeneity. The reference class of days on which there is a local drop in barometric pressure inside my house, for instance, is more homogeneous than the reference class of days on which my barometer shows a sudden drop, for my barometer may be malfunctioning. Similarly, the reference class of days on which there is a widespread sudden drop in atmospheric pressure is more homogeneous than the days on which there is a local drop, for the house may be tightly sealed or the graduate students may be playing a joke on me.[51] It is not that we obtain a large increase in the probability of a storm as we move from one of these reference classes to another; rather, each progressively better partitioning makes the preceding partitioning *statistically irrelevant.*

It will be recalled that the property C is statistically irrelevant to the attribute B in the reference class A iff $P(A,B) = P(A.C,B)$. The probability of a storm on a day when there is a sudden drop in atmospheric pressure and when my barometer executes a sudden drop is precisely the same as the probability of a storm on a day when there is a sudden widespread drop in atmospheric pressure. To borrow a useful notion from Reichenbach, we may say that the sudden widespread drop in atmospheric pressure *screens off* the drop in barometer reading from the occurrence of the storm.[52] The converse relation does not hold. The probability of a storm on a day when the reading on my barometer makes a sudden drop is not equal to the probability of a storm on a day when the reading on my barometer makes a sudden drop and there is a sudden widespread drop in the atmospheric pressure. The sudden drop in barometric reading does not screen off the sudden widespread drop in atmospheric pressure from the occurrence of the storm.

More formally, we may say that D screens off C from B in reference class A iff (if and only if)

$$P(A.C.D,B) = P(A.D,B) \neq P(A.C,B).$$

For purposes of the foregoing example, let $A =$ the class of days in the vicinity of my house, let $B =$ the class of days on which there is an occurrence of a storm, let $C =$ the class of days on which there is a sudden drop in reading on my barometer, and let $D =$ the class of days on which there is a widespread drop in atmospheric pressure in the area in which my house is located. By means of this formal definition, we see that D screens off C from B, but C does not screen off D from B. The screening-off relation is, therefore, not symmetrical, although the relation of statistical relevance is symmetrical.[53]

When one property in terms of which a statistically relevant partition in a reference class can be effected screens off another property in terms of which another statistically relevant partition of that same reference class can be effected, then the screened-off property must give way to the property which screens it off. This is the *screening-off rule*. The screened-off property then becomes irrelevant and no longer has explanatory value. This consideration shows how we can handle the barometer example and a host of others, such as the explanation of measles in terms of spots, in terms of exposure to someone who has the disease, and in terms of the presence of the virus. The unwanted "symptomatic explanations" can be blocked by use of the screening-off concept, which is defined in terms of statistical irrelevance alone. We have not found it necesary to introduce an independent concept of causal relation in order to handle this problem. Reichenbach believed it was possible to define causal relatedness in terms of screening-off relations; but whether his program can be carried through or not, it seems that many causal relations exhibit the desired screening-off relations.[54]

8. Explanations with Low Weight

According to Hempel, the basic requirement for an inductive explanation is that the posterior weight (as I have been describing it) must be high, whereas I have been suggesting that the important characteristic is the increase of the posterior weight over the prior weight as a result of incorporating the event into a homogeneous reference class. The examples discussed thus far satisfy both of these desiderata, so they do not serve well to discriminate between the two views. I would maintain, however, that when the prior weight of an event is very low, it is not necessary that its posterior weight be made high in order to have an inductive explanation. This point is illustrated by the well-known paresis example.[55]

No one ever contracts paresis unless he has had latent syphilis which has gone untreated, but only a small percentage of victims of untreated latent syphilis develop paresis. Still, it has been claimed, the occurrence of paresis is explained by the fact that the individual has had syphilis. This example has been hotly debated because of its pertinence to the issue of symmetry between explanation and prediction—an issue I still wish to postpone. Nevertheless, the following observations are in order. The prior probability of a person contracting paresis is very low, and the reference class of people in general is inhomogeneous. We can make a statistically relevant partition into people who have untreated latent syphilis and those who do not. (Note that latent syphilis screens off primary syphilis and secondary syphilis.) The probability that a person with untreated latent syphilis will contract paresis is still low, but it is considerably higher than the prior probability of paresis among people in general. To cite untreated latent syphilis as an explanation of paresis is correct, for it does provide a partition of the general reference class which yields a more homogeneous reference class and a higher posterior weight for the explanandum event.

When the posterior weight of an event is low, it is tempting to think that we have not fully explained it. We are apt to feel that we have not yet found a completely homogeneous reference class. If only we had fuller understanding, we often believe, we could sort out causal antecedents in order to be able to say which cases of untreated latent syphilis will become paretic and which ones will not. With this knowledge we would be able to partition the reference class of victims of untreated latent syphilis into two subclasses in terms of these causal antecedents so that all (or an overwhelming majority of) members of one subclass will develop paresis whereas none (or very few) of the members of the other will become paretic. This conviction may be solidly based upon experience with medical explanation, and it may provide a sound empirical basis for the search for additional explanatory facts—more relevant properties in terms of which to improve the homogeneity of the reference class. Nevertheless, the reference class of untreated latent syphilitics is (as I understand it) epistemically homogeneous in terms of our present knowledge, so we have provided the most adequate explanation possible in view of the knowledge we possess.[56]

A parallel example could be constructed in physics where we have much greater confidence in the actual homogeneity of the reference class. Suppose we had a metallic substance in which one of the atoms experienced radioactive decay within a particular, small time period, say one minute. For purposes of the example, let us suppose that only one such decay occurred within that time period. When asked why that particular atom decayed, we might reply that the substance is actually an alloy of two metals, one radioactive (for example, uranium 238) and one stable (for example, lead 206). Since the half-life of U^{238} is 4.5×10^9 years, the probability of a given uranium atom's decaying in an interval of one minute is not large, yet there is explanatory relevance in pointing out that the atom that did decay was a U^{238} atom.[57] According to the best theoretical knowledge now available, the class of U^{238} atoms is homogeneous with respect to radioactive decay, and there is in principle no further relevant partition that can be made. Thus, it is not necessary to suppose that examples such as the paresis case derive their explanatory value solely from the conviction that they are partial explanations which can someday be made into full explanations by means of further knowledge.[58]

There is one further way to maintain that explanations of improbable events involve, nevertheless, high probabilities. If an outcome is improbable (though not impossible) on a given trial, its occurrence at least once in a sufficiently large number of trials can be highly probable. For example, the probability of getting a double six on a toss of a pair of standard dice is 1/36; in twenty-five tosses the probability of at least one double six is over one-half. No matter how small the probability p of an event on a single trial, provided $p > 0$, and no matter how large r, provided $r < 1$, there is some n such that the probability of at least one occurrence in n trials is greater than r.[59] On this basis, it might be claimed, we explain improbable events by saying, in effect, that given enough trials the event is probable. This is a satisfactory explanation of the fact that certain types of events occur occasionally, but it still leaves open the question of how to explain the fact that this partcuilar improbable event happened on this particular occasion.

To take a somewhat more dramatic example, each time an alpha particle bombards the potential barrier of a uranium nucleus, it has a chance of about 10^{-38} of tunneling through and escaping from the nucleus; one can appreciate the magnitude of this number by noting that whereas the alpha particle bombards the potential barrier about 10^{21} times per second, the half-life of uranium is of the order of a billion years.[60] For any given uranium atom, if we wait long enough, there is an overwhelmingly large probability that it will decay, but if we ask, "Why did the alpha particle tunnel out on this particular bombardment of the potential barrier?" the only answer is that in the homogeneous reference class of approaches to the barrier, it has a 1-in-10^{38} chance of getting through. I do not regard the fact that it gets through on a particular trial inexplicable, but certainly anyone who takes explanations to be arguments showing that the event was to be expected must conclude that this fact defies all explanation.

9. Multiple Homogeneity

The paresis example illustrates another important methodological point. I have spoken so far as if one homogeneous reference class were sufficient for the explanation of a particular event. This, I think, is incorrect. When a general reference class is partitioned, we can meaningfully ask about the homogeneity of each subclass in the partition (as we did in defining degree of inhomogeneity above). To be sure, when we are attempting to provide an explanation of a particular explanandum event x, we focus primary attention upon the subclass to which x belongs. Nevertheless, I think we properly raise the question of the homogeneity of other subclasses when we evaluate our explanation. In the paresis example we may be convinced that the reference class of untreated latent syphilitics is inhomogeneous, but the complementary class is perfectly homogeneous. Since no individuals who do not have untreated latent syphilis develop paresis, no partition statistically relevant to the development of paresis can be made in the reference class of people who do not have untreated latent syphilis.

Consider the table salt example from the standpoint of the homogeneity of more than one reference class. Although the reference class of samples of table salt is completely homogeneous for water solubility, the complementary class certainly is not. It is possible to make further partitions in the class of samples of substances other than table salt which are statistically relevant to water solubility: samples of sand, wood, and gold are never water soluble; samples of baking soda, sugar, and rock salt always are.

If we explain the fact that this sample dissolves in water by observing that it is table salt and all table salt is water soluble, we may feel that the explanation is somewhat inadequate. Some theorists would say that it is an adequate explanation, but that we can equally legitimately ask for an explanation of the general fact that table salt is water soluble. Although I have great sympathy with the idea that general facts need explanation and are amenable to explanation, I think it is important to recognize the desideratum of homogeneity of the complementary reference class. I think we may rightly claim fully adequate explanation of a particular fact when (but not necessarily only when) the original reference class A, with respect to which its prior probability is assessed, can be partitioned into two subclasses $A.C$ and $A.\bar{C}$, each of which is homogeneous for the attribute in question. In the ideal case all $A.C$'s are B and no $A.\bar{C}$'s are B—that is, if x is A, then x is B if and only if x is C. However, there is no reason to believe that Nature is so accommodating as to provide in all cases even the possibility in principle of such fully deterministic explanations.

We now have further reason to reject the dissolving spell explanation for the fact that a sample of table salt dissolves in water. If we partition the general reference class of samples of unspecified substances into hexed samples of table salt and all other samples of substances, this latter reference class is less homogeneous than the class of all samples of substances other than table salt, for we know that all unhexed samples of table salt are water soluble. To make a statistically irrelevant partition not only reduces the available statistics; it also reduces the homogeneity of the complementary reference class. This consideration also applies to such other examples as John Jones and his wife's birth control pills.

It would be a mistake to suppose that it must always be possible in principle to partition the original reference class A into two homogeneous subclasses. It may be that the best we can hope for is a partition into k subclasses $A.C_k$, each completely homogeneous, and such that $P(A.C_i,B) \neq P(A.C_j,B)$ if $i \neq j$. This is the *multiple homogeneity rule.* It expresses the fundamental condition for adequate explanation of particular events, and it will serve as a basis for the general characterization of deductive and inductive explanation.

The multiple homogeneity requirement provides, I believe, a new way to attack a type of explanation that has been extremely recalcitrant in the face of approaches similar to Hempel's, namely, the so-called *functional explanation.* We are told, for example, that it is the function of hemoglobin to transport oxygen from the lungs to the cells in the various parts of the organism. This fact is taken to provide an explanation for the blood's hemoglobin content. Trouble arises when we try to fit such explanations to schemata like (3) for deductive explanation, for we always seem to end up with a necessary condition where we want a sufficient condition. Consider the following:

(11) Hemoglobin transports oxygen from the lungs to the other parts of the body (in an animal with a normally functioning circulatory system).
This animal has hemoglobin in its blood.
___
This animal has oxygen transported from its lungs to the other parts of its body.

Here we have a valid deductive argument, but unfortunately the explanandum is a premise and the conclusion is part of the explanans. This will never do. Let us interchange the second premise and the conclusion, so that the explanandum becomes the conclusion and the explanans consists of all of the premises, as follows:

(12) Hemoglobin transports oxygen from the lungs to the other parts of the body.
This animal has oxygen transported from its lungs to other parts of its body.
___
This animal has hemoglobin in its blood.

Now we get an obviously invalid argument. In order to have a standard deductive explanation of the Hempel variety, the particular explanatory fact cited in the explanans must, in the presence of the general laws in the explanans, be a sufficient condition for the explanandum event. In the hemoglobin case the explanatory fact is a necessary condition of the explanandum event, and that is the typical situation with functional explanations.

Although the suggestion that we admit that necessary conditions can have explanatory import might seem natural at this point, it meets with an immediate difficulty. We would all agree that the fact that a boy was born does not contribute in the least to the explanation of his becoming a juvenile delinquent, but it certainly was a necessary condition of that occurrence. Thus, in general necessary conditions cannot be taken to have any explanatory function, but it is still open to us to see whether under certain special conditions necessary conditions can function in genuine explanations. The approach I am suggesting enables us to do just that.

Returning to the paresis example, suppose that medical science could discover in the victims of untreated latent syphilis some characteristic *P* that is present in each victim who develops paresis and is lacking in each one who escapes paresis. Then, in the reference class S of untreated latent syphilitics, we could make a partition in which every member of one subclass would develop paresis, *B*, whereas no member of the other subclass would fall victim to it. In such a case, within S, *P* would be a necessary and sufficient condition for *B*. If, however, medical research were slightly less successful and found a characteristic *Q* that is present in almost every member of S who develops paresis, but is lacking in almost every one who escapes paresis, then we would still have a highly relevant partition of the original reference class into two subclasses, and this would be the statistical analogue of the discovery of necessary and sufficient conditions. Thus, if

$$P(S,B) = p_1;\ P(S.Q,B) = p_2;\ P(S.\bar{Q},B) = p_3$$

where

$$p_1 \neq p_2 \text{ and } p_1 \neq p_3,$$

then *Q* is a statistically relevant statistical analogue of a necessary condition. To concern ourselves with the homogeneity of $S.\bar{Q}$ is to raise the question of further statistically relevant necessary conditions.

I have argued above that sufficient conditions have explanatory import only if they are relevant. Insertion of a sodium salt into a Bunsen flame is a relevant sufficient condition of the flame turning yellow; taking birth control pills is an irrelevant sufficient condition for nonpregnancy in a man. Similarly, being born is an irrelevant necessary condition for a boy becoming delinquent, for it makes no partition whatever in the reference class of boys, let alone a relevant one. However, as argued above, latent untreated syphilis is a relevant necessary condition of paresis, and it has genuine explanatory import.

Let us see whether such considerations are of any help in dealing with functional explanations. According to classical Freudian theory, dreams are functionally explained as wish fulfillments. For instance, one night, after a day of fairly frustrating work on a review of a book containing an example about dogs who manifest the Absolute by barking loudly, I dreamed that I was being chased by a pack of barking dogs. Presumably, my dream had the function of enabling me to fulfill my wish to escape from the problem of the previous day, and thus to help me preserve my sleep. The psychoanalytic explanation does not maintain that only one particular dream would have fulfilled this function; presumably many others would have done equally well. Thus, the fact that I had that particular wish did not entail (or even make it very probable) that I would have that particular dream, but the fact that I had the dream was, allegedly, a sufficient condition for the fulfilment of the wish. The explanatory fact, the unfulfilled wish, is therefore taken to be a necessary condition of the explanandum event—the dream—but not a sufficient condition. According to the psychoanalytic theory, the unfulfilled wish is a relevant necessary condition, however, for if we make a partition of the general reference class of nights of sleep in terms of the events of the preceding day, then the probability of this particular dream is much greater in the narrower reference class than it was in the original reference class.[61] By the way, the author of the book I was reviewing is Stephen Barker.

When we try to give a functional explanation of a particular mechanism, such as the presence of hemoglobin or the occurrence of a dream, we may feel some dissatisfaction if we are unable eventually to explain why that particular mechanism occurred rather than another which would apparently have done the same job equally well. Such a feeling is, I would think, based upon the same considerations that lead us to feel dissatisfied with the explanation of paresis. We know some necessary conditions, in all of these cases, but we believe that it is possible in principle to find sufficient conditions as well, and we will not regard the explanations as complete until we find them. How well this feeling is justified depends chiefly upon the field of science in which the explanation is sought, as the example of radioactive decay of a U^{238} atom shows.

The fact that the multiple homogeneity rule allows for the explanatory value of relevant necessary conditions and their statistical analogues does not, by itself, provide a complete account of functional explanation, for much remains to be said about the kinds of necessary conditions that enter into functional explanations and their differences from necessary conditions that are not in any sense functional. Our approach has, however, removed what has seemed to many authors the chief stumbling block in the path to a satisfactory theory of functional explanation.

10. Explanation Without Increase of Weight

It is tempting to suppose, as I have been doing so far and as all the examples have suggested, that explanation of an explanandum event somehow confers upon it a posterior weight that is greater than its prior weight. When I first enunciated this principle in section 3, however, I indicated that, though heuristically beneficial, it should not be considered fully accurate. Although most explanations may conform to this general principle, I think there may be some that do not. It is now time to consider some apparent exceptions and to see whether they are genuine exceptions.

Suppose, for instance, that a game of heads and tails is being played with two crooked pennies, and that these pennies are brought in and out of play in some irregular manner. Let one penny be biased for heads to the extent that 90 percent of the tosses with it yield heads; let the other be similarly biased for tails. Furthermore, let the two pennies be used with equal frequency in this game, so that the overall probability of heads is one-half. (Perhaps a third penny, which is fair, is tossed to decide which of the two biased pennies is to be used for any given play.) Suppose a play of this game results in a head; the prior weight of this event is one-half. The general reference class of plays can, however, be partitioned in a statistically relevant way into two homogeneous reference classes. If the toss was made with the penny biased for heads, the result is explained by that fact, and the weight of the explanandum event is raised from 0.5 to 0.9.

Suppose, however, that the toss were made with the penny biased for tails; the explanandum event is now referred to the other subclass of the original reference class, and its weight is decreased from 0.5 to 0.1. Do we want to say in this case that the event is thereby explained? Many people would want to deny it, for such cases conflict with their intuitions (which, in many cases, have been significantly conditioned by Hempel's persuasive treatment) about what an explanation ought to be. I am inclined, on the contrary, to claim that this is genuine explanation. There are, after all, improbable occurrences—such events are not explained by making them probable. Is it not a peculiar prejudice to maintain that only those events which are highly probable are capable of being explained—that improbable events are in principle inexplicable? Any event, regardless of its probability, is amenable to explanation, I believe; in the case of improbable events, the correct explanation is that they are highly improbable occurrences which happen, nevertheless, with a certain definite frequency. If the reference class is actually homogeneous, there are no other circumstances with respect to which they are probable. No further explanation can be required or can be given.[62]

There are various reasons for which my view might be rejected. In the first place, I am inclined to think that the deterministic prejudice may often be operative—namely, that x is B is not explained until x is incorporated within a reference class all of whose members are B. This is the feeling that seemed compelling in connection with the paresis example. In the discussion of that example, I indicated my reasons for suggesting that our deterministic hopes may simply be impossible to satisfy. In an attempt to undercut the hope generated in medical science that determinism would eventually triumph, I introduced the parallel example of the explanation of a radioactive decay in an alloy of lead and uranium, in which case there is strong reason to believe that the reference class of uranium 238 atoms is strictly homogeneous with respect to disintegration.

In order to avoid being victimized by the same deterministic hope in the present context, we could replace the coin-tossing example at the beginning of this section with another example from atomic physics. For instance, we could consider a mixture of uranium 238 atoms, whose half-life is 4.5×10^9 years, and polonium 214 atoms, whose half-life is 1.6×10^{-4} seconds.[63] The probability of disintegration of an unspecified atom in the mixture is between that for atoms of U^{238} and Po^{214}. Suppose that within some small specified time interval a decay occurs. There is a high probability of a polonium atom disintegrating within that interval, but a very low probability for a uranium atom. Nevertheless, a given disintegration may be of a uranium atom, so the transition from the reference class of a mixture of atoms of the two types to a reference class of atoms of U^{238} may result in a considerable lowering of the weight. Nevertheless, the latter reference class may be unqualifiedly homogeneous. When we ask why that particular atom disintegrated, the answer is that it was a U^{238} atom, and there is a small probability that such an atom will disintegrate in a short time interval.

If, in the light of modern developments in physics and philosophy, determinism no longer seems tenable as an a priori principle, we may try to salvage what we can by demanding that an explanation that does not necessitate its explanandum must at least make it highly probable. This is what Hempel's account requires. I have argued above, in the light of various examples, that even this demand is excessive and that we must accept explanations in which the explanandum event ends up with a low posterior weight. "Well, then," someone might say, "if the explanation does not show us that the event was to be expected, at least it ought to show us that the event was to be expected somewhat more than it otherwise would have been." But this attitude seems to derive in an unfortunate way from regarding explanations as arguments. At this juncture it is crucial to point out that the emphasis in the present account of explanation is upon achieving a relevant partition of an inhomogeneous reference class into homogeneous subclasses. On this conception an explanation is not an argument that is intended to produce conviction; instead, it is an attempt to assemble the factors that are relevant to the occurrence of an event. There is no more reason to suppose that such a process will increase the weight we attach to such an occurrence than there is to suppose that it will lower it. Whether the posterior weight is higher than or lower than the prior weight is really beside the point. I shall have more to say later about the function of explanations.

Before leaving this topic, I must consider one more tempting principle. It may seem evident from the examples thus far considered that an explanation must result in a change—an increase or a decrease—in the transition from the prior weight to the posterior weight. Even this need not occur. Suppose, for instance, that we change the coin-tossing game mentioned at the outset of this section by introducing a fair penny into the play. Now there are three pennies brought into play randomly: one with a probability of 0.9 for heads, one with a probability of 0.5 for heads, and one with a probability of 0.1 for heads. Overall, the probability of heads in the game is still one-half. Now, if we attempt to explain a given instance of a head coming up, we may partition the original reference class into three homogeneous subclasses, but in one of these three the probability of heads is precisely the same as it is in the entire original class. Suppose our particular head happens to belong to that subclass. Then its prior weight is exactly the same as its posterior weight, but I would claim that explanation has occurred simply by virtue of the relevant partition of the original nonhomogeneous reference class. This makes sense if one does not insist upon regarding explanations as arguments.

11. Some Paradigms of Explanation

Before attempting a general quasi-formal characterization of the explanation of particular events, I should like to follow a common practice and introduce some examples that seem to me to deserve the status of paradigms of explanation. They will, I believe, differ in many ways from the paradigms that are usually offered; they come from a set of investigations, conducted mainly by Grünbaum and Reichenbach, concerning the temporal asymmetry (or anisotropy) of the physical world.[64] Because of the close connections between causality and time on the one hand, and between causality and explanation on the other, these investigations have done a great deal to elucidate the problem of explanation.

Given a thermally isolated system, there is a small but nonvanishing probability that it will be found in a low entropy state. A permanently closed system will from time to time, but very infrequently, spontaneously move from a state of high entropy to one of low entropy, although by far the vast majority of its history is spent in states of high entropy. Let us take this small probability as the prior probability that a closed system will be in a low entropy state. Suppose, now, that we examine such a system and find that it is in a low entropy state. What is the explanation? Investigation reveals that the system, though closed at the time in question, had recently interacted with its environment. The low entropy state is explained by this interaction. The low entropy of the smaller system is purchased by an expenditure of energy in the larger environmental system that contains it and by a consequent increase in the entropy of the environment.

For example, there is a small but nonvanishing probability that an ice cube will form spontaneously in a thermos of cool water. Such an occurrence would be the result of a chance clustering of relatively nonenergetic molecules in one particular place in the container. There is a much greater chance that an ice cube will be present in a thermos of cool water to which an ice cube has just been added. Even if we had no independent knowledge about an ice cube having been added, we would confidently infer it from the mere presence of the ice cube. The low entropy state is explained by a previous interaction with the outside world, including a refrigerator that manufactures the ice cubes.

Suppose, for a further example, that we found a container with two compartments connected by an opening, one compartment containing only oxygen and the other only nitrogen. In this case, we could confidently infer that the two parts of the container had been separately filled with oxygen and nitrogen and that the connecting window had recently been opened—so recently that no diffusion had yet taken place. We would not conclude that the random motions of the molecules had chanced to separate the two gases into the two compartments, even though that event has a small but nonvanishing probability. This otherwise improbable state of affairs is likewise explained in terms of a recent interaction with the environment.

In examples of these kinds, the antecedent reference class of states of closed systems is recognized as inhomogeneous, and it is partitioned into those states which follow closely upon an interaction with the outside world and those that do not. Within the former homogeneous subclass of the original reference class, the probability of low entropy states is much higher than the probability of such states referred to the unpartitioned inhomogeneous reference class. In such cases we, therefore, have explanations of the low entropy states.

The foregoing examples involve what Reichenbach calls "branch systems." In his extended and systematic treatment of the temporal asymmetry of the physical world, he accepts it as a fundamental fact that our region of the universe has an abundant supply of these branch systems. During their existence as separate systems they may be considered closed, but they have not always been isolated, for each has a definite beginning to its history as a closed system. If we take one such system and suppose it to exist from then on as a closed system, we can consider its successive entropy states as a probability sequence, and the probability of low entropy in such a sequence is very low. This probability is a *one-system probability* referred to a *time ensemble;* as before, let us take it as a prior probability. If we examine many such systems, we find that a large percentage of them are in low entropy states shortly after their inceptions as closed systems. There is, therefore, a much higher probability that a very young branch system is in a low entropy state. This is the posterior probability of a low entropy state for a system that has recently interacted with its environment; it is a *many-system probability* referred to a *space ensemble*. The time ensemble turns out to be an inhomogeneous reference class; the space ensemble yields homogeneous reference classes.[65] Applying this consideration to our ice cube example, we see that the presence of the ice cube in the thermos of cool water is explained by replacing the prior weight it would have received from the reference class of states of this system as an indefinitely isolated container of water with the weight it receives from the reference class of postinteraction states of many such containers of water when the interaction involves the placing of an ice cube within it.

Even if we recognize that the branch systems have not always existed as closed systems, it is obviously unrealistic to suppose that they will remain forever isolated. Instead, each branch system exists for a finite time, and at each temporal end it merges with its physical environment. Each system exhibits low entropy at one temporal end and high entropy at the other. The basic fact about the branch systems is that the vast majority of them exhibit low entropy states at the same temporal end. We can state this fact without making any commitment whatever as to whether that end is the earlier or the later end. However, the fact that the systems are alike in having low entropy at the same end can be used to provide coordinating definitions of such temporal relations as *earlier* and *later*. We arrive at the usual temporal nomenclature if we say that the temporal end at which the vast majority of branch systems have low entropy is the earlier end and the end at which most have high entropy is the later end. It follows, of course, that the interaction with the environment at the low entropy end precedes the low entropy state.

Reichenbach takes this relationship between the initial interaction of the branch system with its environment and the early low entropy state to be the most fundamental sort of *producing*.[66] The interaction produces the low entropy state, and the low entropy state is the product (and sometimes the record) of the interaction. Producing is, of course, a thoroughly causal concept; thus, the relation of interaction to orderly low entropy state becomes the paradigm of causation on Reichenbach's view, and the explanation of the ice cube in the thermos of water is an example of the most fundamental kind of causal explanation. The fact that this particular interaction involves a human act is utterly immaterial, for many other examples exist in which human agency is entirely absent.

Reichenbach's account, if correct, shows why causal explanation is temporally asymmetrical. Orderly states are explained in terms of previous interactions; the interactions that are associated with low entropy states do not follow them. It appears that the temporal asymmetry of causal explanation is preserved even when the causal concepts are extended to refer to reversible mechanical processes. It is for this reason, I imagine, that we are willing to accept an explanation of an eclipse in terms of laws of motion and *antecedent* initial conditions, but we feel queasy, to say the least, about claiming that the same eclipse can be *explained* in terms of the laws of motion and *subsequent* initial conditions, even though we may *infer* the occurrence of the eclipse equally well from either set of initial conditions. I shall return to this question of temporal asymmetry of explanation in the next section.

The foregoing examples are fundamentally microstatistical. Reichenbach has urged that macrostatistical occurrences can be treated analogously.[67] For instance, a highly ordered arrangement in a deck of cards can be explained by the fact that it is newly opened and was arranged at the factory. Decks of cards occasionally get shuffled into highly ordered arrangements—just as closed thermodynamic systems occasionally develop low entropy states—but much more frequently the order arises from a deliberate arrangement. Reichenbach likewise argues that common causes macrostatistically explain improbable "coincidences." For example, suppose all the lights in a certain block go off simultaneously. There is a small probability that in every house the people decided to go to bed at precisely the same instant. There is a small probability that all the bulbs burned out simultaneously. There is a much greater probability of trouble at the power company or in the transmission lines. When investigation reveals that a line was downed by heavy winds, we have an explanation of a coincidence with an extremely low prior weight. A higher posterior weight results when the coincidence is assigned to a homogeneous reference class. The same considerations apply, but with much more force, when it is the entire eastern seaboard of the United States that suffers a blackout.

Reichenbach believed that it is possible to establish temporal asymmetry on the basis of macrostatistical relations in a manner quite analogous to the way in which he attempted to base it upon microstatistical facts. His point of departure for this argument is the fact that certain improbable coincidences occur more often than would be expected on the basis of chance alone. Take, as an example, all members of a theatrical company falling victim to a gastrointestinal illness on the same evening. For each individual there is a certain probability (which may, of course, vary from one person to another) that he will become ill on a given day. If the illnesses of the members of the company are independent of one another, the probability of all members of the company becoming ill on the same day is simply the product of all the separate probabilities; for the general multiplication rule says (for two events) that

$$P(A,B.C) = P(A,B) \times P(A.B,C), \tag{1}$$

but if

$$P(A,C) = P(A.B,C), \tag{2}$$

we have the special multiplication rule,

$$P(A,B.C) = P(A,B) \times P(A,C). \tag{3}$$

When relation (3) holds, we say that the events B and C are independent of each other; this is equivalent to relation (2), which is our definition of statistical irrelevance.

We find, as a matter of fact, that whole companies fall ill more frequently than would be indicated if the separate illnesses were all statistically independent of one another. Under these circumstances, Reichenbach maintains, we need a causal explanation for what appears to be an improbable occurrence. In other words, we need a causal explanation for the statistical relevance of the illnesses of the members of the company to one another. The causal explanation may lie in the food that all of the members of the company ate at lunch. Thus, we do not assert a direct causal relation between the illness of the leading man and that of the leading lady; it was not the fact that the leading man got sick that caused the illness of the leading lady (or the actors with minor roles, the stand-ins, the prop men, etc.). There is a causal relation, but it is via a common cause—the spoiled food. The common meal screens off the statistical relevance of the illness of the leading man to that of the leading lady. The case is parallel to that of the barometer and the storm, and the screening off works in the same way.

Reichenbach maintains that coincidences of this sort can be explained by common causes. The illness of the whole company is explained by the common meal; the blackout of an entire region is explained by trouble at the power source or in the distribution lines. If we take account of the probability of spoiled food's being served to an entire group, the frequency with which the members all become victims of a gastrointestinal illness on the same day is not excessive. If we take into account the frequency with which power lines go down or of trouble at the distribution station, etc., the frequency with which whole areas simultaneously go dark is not excessive.

Such occurrences are not explained in terms of common effects. We do not seriously claim that all members of the company became ill because events were conspiring (with or without conscious intent of some agency, natural or supernatural) to bring about a cancellation of the performance; we do not believe that the future event explains the present illness of the company. Similarly, we do not explain the blackout of the eastern seaboard in terms of events conspiring to make executives impregnate their secretaries. The situation is this: in the absence of a common cause, such as spoiled food, the cancellation of the play does not make the coincidental illness of the company any more probable than it would have been as a product of probabilities of independent events.

By contrast, the presence of a common cause such as the common meal does make the simultaneous illness more probable than it would have been as a product of independent events, and that is true whether or not the common effect occurs. The efficacy of the common cause is not affected by the question of whether the play can go on, with a sick cast or with substitutes. The probability of all the lights going out simultaneously, in the absence of a common cause, is unaffected by the question of how the men and women, finding themselves together in the dark for the entire night without spouses, behave themselves. The net result is that coincidences of the foregoing type are to be explained in terms of causal antecedents, not subsequent events. The way to achieve a relevant subdivision of the general reference class is in terms of antecedent conditions, not later ones. This seems to be a pervasive fact about our macroworld, just as it seems to characterize the microworld.

I believe that an important feature of the characterization of explanation I am offering is that it is hospitable to paradigms of the two kinds I have discussed in this section, the microstatistical examples and the macrostatistical examples. In all these cases, there is a prior probability of an event which we recognize as furnishing an inappropriate weight for the event in question. We see that the reference class for the prior probability is inhomogeneous and that we can make a relevant partition. The posterior probability which arises from the new homogeneous reference class is a suitable weight to attach to the single event. It is the introduction of statistically relevant factors for partitioning the reference class that constitutes the heart of these explanations. If these examples are, indeed, paradigms of the most fundamental types of explanation, it is a virtue of the present account that it handles them easily and naturally.

12. The Temporal Asymmetry of Explanation

It is also a virtue of the present account, I believe, that it seems to accommodate the rather puzzling temporal asymmetry of explanation already noted above—the fact that we seem to insist upon explaining events in terms of earlier rather than later initial conditions. Having noted that both the microstatistical approach and the macrostatistical approach yield a fundamental temporal asymmetry, let us now use that fact to deal with a familiar example. I shall attempt to show how Reichenbach's principle of the common cause, introduced in connection with the macrostatistical examples discussed above, helps us to establish the temporal asymmetry of explanation.

Consider Silvain Bromberger's flagpole example, which goes as follows.[68] On a sunny day a flagpole of a particular height casts a shadow of some particular length depending upon the elevation of the sun in the sky. We all agree that the position of the sun and the height of the flagpole explain the length of the shadow. Given the length of the shadow, however, and the position of the sun in the sky, we can equally infer the height of the flagpole, but we rebel at the notion that the length of the shadow explains the height of the flagpole. It seems to me that the temporal relations are crucial in this example. Although the sun, flagpole, and shadow are perhaps commonsensically regarded as simultaneous, a more sophisticated analysis shows that physical processes going on in time are involved. Photons are emitted by the sun, they travel to the vicinity of the flagpole where some are absorbed and some are not, and those which are not go on to illuminate the ground. A region of the ground is not illuminated, however, because the photons traveling toward that region were absorbed by the flagpole. Clearly the interaction between the photons and the flagpole temporally precedes the interaction between the neighboring photons and the ground. The reason that the explanation of the length of the shadow in terms of the height of the flagpole is acceptable, whereas the "explanation" of the height of the flagpole in terms of the length of the shadow is not acceptable, seems to me to hinge directly upon the fact that there are causal processes with earlier and later temporal stages. It takes only a very moderate extension of Reichenbach's terminology to conclude that the flagpole produces the shadow in a sense in which the shadow certainly does not produce the flagpole.

If we give up the notion that explanations are arguments, there is no need to be embarrassed by the fact that we can oftentimes infer earlier events from later ones by means of nomological relations, as in the cases of the eclipse and the flagpole. I have been arguing that relevance considerations are preeminent for explanations; let us see whether this approach provides a satisfactory account of the asymmetry of explanation in the flagpole case. The apparent source of difficulty is that, under the general conditions of the example, the flagpole is relevant to the shadow and the shadow is relevant to the flagpole. It does not follow, however, that the shadow explains the flagpole as well as the flagpole explains the shadow.

In order to analyze the relevance relations more carefully, I shall reexamine a simplified version of the example of the illness in the theatrical company and compare it with a simplified version of the flagpole example. In both cases the screening-off relation will be employed in an attempt to establish the temporal asymmetry of the explanation. The general strategy will be to show that a common cause screens off a common effect and, consequently, by using the screening-off rule, that the explanation must be given in terms of the common cause and not in terms of the common effect. In order to carry out this plan, I shall regard the flagpole as an orderly arrangement of parts that requires an explanation of the same sort as does the coincidental illnesses of the actors.

Consider, then, the simple case of a theatrical company consisting of only two people, the leading lady and the leading man. Let A be our general reference class of days, let M be the illness of the leading man, let L be the illness of the leading lady, let F be a meal of spoiled food that both eat, and let C be the cancellation of the performance. Our previous discussion of the example has shown that the simultaneous illness occurs more frequently than it would if the two were independent, that is,

$$P(A,L.M) > P(A,L) \times P(A,M),$$

and that the common meal is highly relevant to the concurrent illnesses, that is,

$$P(A.F,L.M) > P(A,L.M) > P(A.\bar{F},L.M).$$

It is not true, without further qualification, that the eating of spoiled food F screens off the cancellation of the performance C from the joint illness $L.M$. Although the eating of spoiled food does make it probable that the two actors will be ill, the fact that the performance has been cancelled supplies further evidence of the illness and makes the joint illness more probable. At this point experiment must be admitted. It is clearly possible to arrange things so that the play goes on, illness or no illness, by providing substitutes who never eat with the regular cast. Likewise, it is a simple matter to see to it that the performance is cancelled, whether or not anyone is ill. Under these experimental conditions we can see that the probability of the joint illness, given the common meal with spoiled food, is the same whether or not the play is performed, that is,

$$P(A.F.C,L.M) = P(A.F.\bar{C},L.M).$$

From this it follows that

$$P(A.F.C,L.M) = P(A.F,L.M).$$

If we approach the common meal of spoiled food in the same experimental fashion, we can easily establish that it is not screened off by the cancellation of the performance. We can arrange for the two leading actors to have no meals supplied from the same source and ascertain the probability $P(A.\bar{F}.C,L.M)$. This can be compared with the probability $P(A.F.C,L.M)$ which arises under the usual conditions of the two actors eating together in the same restaurants. Since they are not equal, the common spoiled food is statistically relevant to the joint illness, even in the presence of the cancellation of the performance, that is,

$$P(A.F.C,L.M) \neq P(A.C,L.M).$$

Thus, although the common cause F is relevant to the coincidence $L.M$ and the coincidence $L.M$ is relevant to C (from which it follows that C is relevant to $L.M$), it turns out that the common cause F screens off the common effect C from the coincidence to be explained. By the screening-off rule, F must be used and C must not be used to partition the reference class A. These considerations express somewhat formally the fact that tampering with the frequency of F without changing the frequency of C will affect the frequency of $L.M$, but tampering with the frequency of C without changing the frequency of F will have no effect on the frequency of $L.M$. This seems to capture the idea that we can influence events by influencing their causal antecedents, but not by influencing their causal consequents.

Let us apply the same analysis to the flagpole example. Again, let A be the general reference class, and let us for simplicity suppose that the flagpole is composed of two parts, a top and a bottom. The flagpole is in place when the two parts are in place; let T be the proper positioning of the top, and B the proper positioning of the bottom. Let M represent the flagpole makers' bringing the pieces together in the appropriate positions. Let S be the occurrence of the full shadow of the flagpole. Now, since the flagpole's existence consists in the two pieces' being put together in place and since that hardly happens except when the flagpole makers bring them together and assemble them, it is clear that

$$P(A,T.B) > P(A,T) \times P(A,B).$$

Hence, the existence of the flagpole is something to be explained. Since we know that

$$P(A.M,T.B) > P(A,T.B) > P(A.\bar{M},T.B),$$

M effects a relevant partition in the general reference class A.

The question of whether the flagpole gets put together when the flagpole makers go about putting it together is unaffected by the existence or nonexistence of a shadow—for example, the ground might be illuminated by other sources of light besides the sun, or mirrors might deflect some of the sun's rays, with the result that there is no shadow—but the flagpole is there just the same; hence,

$$P(A.M.S,T.B) = P(A.M,T.B),$$

but

$$P(A.M.S,T.B) \neq P(A.S,T.B),$$

since the shadow can easily be obliterated at will or produced by other means without affecting the existence of the flagpole. We, therefore, conclude again that the common cause screens off the common effect and that, by virtue of the screening-off rule, the causal antecedent M—and not the causal consequent S—must be invoked to effect a relevant partition in the reference class A. In this highly schematic way, I hope I have shown that the approach to explanation via statistical relevance has allowed us to establish the temporal asymmetry of explanation. Once more, the intuitive idea is that manipulating the pieces of the flagpole, without otherwise tampering with the shadow, affects the shadow; contrariwise, tampering with the shadow, without otherwise manipulating the pieces of the flagpole, has no effect upon the flagpole. The analysis of the flagpole example may, of course, require the same sort of appeal to experiment as we invoked in the preceding example.

I hope it is obvious from all that has been said that none of us is claiming that the temporal asymmetries herein discussed are anything beyond very general facts about the world. Reichenbach regards it as a matter of fact, not of logical necessity, that common causes have their particular statistical relations to the "coincidences" they explain. In a different universe it might be quite otherwise. Neither Reichenbach nor Grünbaum regards the entropic behavior of branch systems as logical necessities, so they are both prepared to deny any logical necessity to the fact that low entropy states are explained by previous interactions. Again, in a different universe, or in another part of this one, it might be different. Similarly, when I argue that explanations from earlier initial conditions are acceptable, whereas explanations from later initial conditions are inadmissible, I take this to be an expression of certain rather general facts about the world. I certainly do not take this characteristic of explanation to be a matter of logical necessity.

The general fact about the world that seems to be involved is that causal processes very frequently exhibit the following sort of structure: a process leading up to a given event E consists of a series of events earlier than E, but such that later ones screen off earlier ones. In other words, a given antecedent event A_1 will be relevant to E, but it will be screened off by a later antecedent A_2 that intervenes between A_1 and E. This situation obtains until we get to E, and then every subsequent event is screened off by some causal antecedent or other. Thus, in some deeply significant sense, the causal consequents of an event are made irrelevant to its occurrence in a way in which the causal antecedents are not. If, in Velikovsky-like cataclysm, a giant comet should disrupt the solar system, it would have enormous bearing upon subsequent eclipses but none whatever upon previous ones. The working out of the details of the eclipse example, along the lines indicated by the analysis of the flagpole example, is left as an exercise for the reader.

To say that causal antecedents can affect an event in ways in which causal consequents cannot is, of course, an utter banality. The difficulty is that causal consequents are *not* statistically irrelevant to an event, so it is rather problematic to state with any precision the kind of asymmetry of relevance we are trying to capture. Reichenbach, himself, is quite careful to point out that the notion of influencing future events is infected with a fundamental temporal ambiguity.[69] I am not at all confident that the use of the screening-off relation is sufficient for a complete account of the relevance relations we seek, but it does have a promising kind of asymmetry, and so I am led to hope that it will contribute at least part of the answer.

13. The Nature of Statistical Explanation

Let me now, at long last, offer a general characterization of explanations of particular events. As I have suggested earlier, we may think of an explanation as an answer to a question of the form, "Why does this x which is a member of A have the property B?" The answer to such a question consists of a partition of the reference class A into a number of subclasses, all of which are homogeneous with respect to B, along with the probabilities of B within each of these subclasses. In addition, we must say which of the members of the partition contains our particular x. More formally, an explanation of the fact that x, a member of A, is a member of B would go as follows:

$$P(A.C_1,B) = p_1$$
$$P(A.C_2,B) = p_2$$
$$\vdots$$
$$P(A.C_n,B) = p_n$$

where

$A.C_1, A.C_2, \ldots, A.C_n$ is a homogeneous partition of A with respect to B,
$p_i = p_j$ only if $i = j$, and
$x \in A.C_k$.

With Hempel, I regard an explanation as a linguistic entity, namely, a set of statements, but unlike him, I do not regard it as an argument. On my view, an explanation is a set of probability statements, qualified by certain provisos, plus a statement specifying the compartment to which the explanadum event belongs.

The question of whether explanations should be regarded as arguments is, I believe, closely related to the question, raised by Carnap, of whether inductive logic should be thought to contain rules of acceptance (or detachment).[70] Carnap's problem can be seen most clearly in connection with the famous lottery paradox. If inductive logic contains rules of inference which enable us to draw conclusions from premises—much as in deductive logic—then there is presumably some number r which constitutes a lower bound for acceptance. Accordingly, any hypothesis h whose probability on the total available relevant evidence is greater than or equal to r can be accepted on the basis of that evidence. (Of course, h might subsequently have to be rejected on the basis of further evidence.) The problem is to select an appropriate value for r. It seems that no value is satisfactory, for no matter how large r is, provided it is less than one, we can construct a fair lottery with a sufficient number of tickets to be able to say for each ticket that will not win, because the probability of its not winning is greater than r. From this we can conclude that no ticket will win, which contradicts the stipulation that this is a fair lottery—no lottery can be considered fair if there is *no* winning ticket.

It was an exceedingly profound insight on Carnap's part to realize that inductive logic can, to a large extent anyway, dispense entirely with rules of acceptance and inductive inferences in the ordinary sense. Instead, inductive logic attaches numbers to hypotheses, and these numbers are used to make practical decisions. In some circumstances such numbers, the degrees of confirmation, may serve as fair betting quotients to determine the odds for a fair bet on a given hypothesis. There is no rule that tells one when to accept an hypothesis or when to reject it; instead, there is a rule of practical behavior that prescribes that we so act as to maximize our expectation of utility.[71] Hence, inductive logic is simply not concerned with inductive arguments (regarded as entities composed of premises and conclusions).

Now, I do not completely agree with Carnap on the issue of acceptance rules in inductive logic; I believe that inductive logic does require some inductive inferences.[72] But when it comes to probabilities (weights) of single events, I believe that he is entirely correct. In my view, we must establish by inductive inference probability statements, which I regard as statements about limiting frequencies. But, when we come to apply this probability knowledge to single events, we procure a weight which functions just as Carnap has indicated—as a fair betting quotient or as a value to be used in computing an expectation of utility.[73] Consequently, I maintain, in the context of statistical explanation of individual events, we do not need to try to establish the explanandum as the conclusion of an inductive argument; instead, we need to establish the weights that would appropriately attach to such explanandum events for purposes of betting and other practical behavior. That is precisely what the partition of the reference class into homogeneous subclasses achieves: it establishes the correct weight to assign to *any* member of A with respect to its being a B. First, one determines to which compartment C_k it belongs, and then one adopts the value p_k as the weight. Since we adopted the *multiple homogeneity rule,* we can genuinely handle any member of A, not just those which happen to fall into one subclass of the original reference class.

One might ask on what grounds we can claim to have characterized explanation. The answer is this. When an explanation (as herein explicated) has been provided, we know exactly how to regard any A with respect to the property B. We know which ones to bet on, which to bet against, and at what odds. We know precisely what degree of expectation is rational. We know how to face uncertainty about an A's being a B in the most reasonable, practical, and efficient way. We know every factor that is relevant to an A having property B. We know exactly the weight that should have been attached to the prediction that this A will be a B. We know all of the regularities (universal or statistical) that are relevant to our original question. What more could one ask of an explanation?

There are several general remarks that should be added to the foregoing theory of explanation:

a. It is evident that explanations as herein characterized are nomological. For the frequency interpretation probability statements are statistical generalizations, and every explanation must contain at least one such generalization. Since an explanation essentially consists of a set of statistical generalizations, I shall call these explanations "statistical" without qualification, meaning thereby to distinguish them from what Hempel has recently called "inductive-statistical." [74] His inductive-statistical explanations contain statistical generalizations, but they are inductive inferences as well.

b. From the standpoint of the present theory, deductive-nomological explanations are just a special case of statistical explanation. If one takes the frequency theory of probability as literally dealing with infinite classes of events, there is a difference between the universal generalization, "All *A* are *B*," and the statistical generalization, "$P(A,B) = 1$," for the former admits no *A*s that are not *B*s, whereas the latter admits of infinitely many *A*s that are not *B*s. For this reason, if the universal generalization holds, the reference class *A* is homogeneous with respect to *B*, whereas the statistical generalization may be true even if *A* is not homogeneous. Once this important difference is noted, it does not seem necessary to offer a special account of deductive-nomological explanations.

c. The problem of symmetry of explanation and prediction, which is one of the most hotly debated issues in discussions of explanation, is easily answered in the present theory. To explain an event is to provide the best possible grounds we could have had for making predictions concerning it. An explanation does not show that the event was to be expected; it shows what sorts of expectations would have been reasonable and under what circumstances it was to be expected. To explain an event is to show to what degree it was to be expected, and this degree may be translated into practical predictive behavior such as wagering on it. In some cases the explanation will show that the explanandum event was not to be expected, but that does not destroy the symmetry of explanation and prediction. The symmetry consists in the fact that the explanatory facts constitute the fullest possible basis for making a prediction of whether or not the event would occur. To explain an event is not to predict it ex post facto, but a complete explanation does provide complete grounds for rational prediction concerning that event. Thus, the present account of explanation does sustain a thoroughgoing symmetry thesis, and this symmetry is not refuted by explanations having low weights.

d. In characterizing statistical explanation, I have required that the partition of the reference class yield subclasses that are, in fact, homogeneous. I have not settled for practical or epistemic homogeneity. The question of whether actual homogeneity or epistemic homogeneity is demanded is, for my view, analogous to the question of whether the premises of the explanation must be true or highly confirmed for Hempel's view.[75] I have always felt that truth was the appropriate requirement, for I believe Carnap has shown that the concept of truth is harmless enough.[76] However, for those who feel too uncomfortable with the stricter requirement, it would be possible to characterize statistical explanation in terms of epistemic homogeneity instead of actual homogeniety. No fundamental problem about the nature of explanation seems to be involved.

e. This paper has been concerned with the explanation of single events, but from the standpoint of probability theory, there is no significant distinction between a single event and any finite set of events. Thus, the kind of explanation appropriate to a single result of heads on a single toss of a coin would, in principle, be just like the kind of explanation that would be appropriate to a sequence of ten heads on ten consecutive tosses of a coin or to ten heads on ten different coins tossed simultaneously.

f. With Hempel, I believe that generalizations, both universal and statistical, are capable of being explained. Explanations invoke generalizations as parts of the explanans, but these generalizations themselves may need explanation. This does not mean that the explanation of the particular event that employed the generalization is incomplete; it only means that an additional explanation is possible and may be desirable. In some cases it may be possible to explain a statistical generalization by subsuming it under a higher level generalization; a probability may become an instance for a higher level probability. For example, Reichenbach offered an explanation for equiprobability in games of chance, by constructing, in effect, a sequence of probability sequences.[77] Each of the first level sequences is a single case with respect to the second level sequence. To explain generalizations in this manner is simply to repeat, at a higher level, the pattern of explanation we have been discussing. Whether this is or is not the only method of explaining generalizations is, of course, an entirely different question.

g. In the present account of statistical explanation, Hempel's problem of the "nonconjunctiveness of statistical systematization"[78] simply vanishes. This problem arises because in general, according to the multiplication theorem for probabilities, the probability of a conjunction is smaller than that of either conjunct taken alone. Thus, if we have chosen a value r, such that explanations are acceptable only if they confer upon the explanandum an inductive probability of at least r, it is quite possible that each of the two explananda will satisfy that condition, whereas their conjunction fails to do so. Since the characterization of explanation I am offering makes no demands whatever for high probabilities (weights), it has no problem of nonconjunctiveness.

14. Conclusion

Although I am hopeful that the foregoing analysis of statistical explanation of single events solely in terms of statistical relevance relations is of some help in understanding the nature of scientific explanation, I should like to cite, quite explicitly, several respects in which it seems to be incomplete.

First, and most obviously, whatever the merits of the present account, no reason has been offered for supposing the type of explanation under consideration to be the only legitimate kind of scientific explanation. If we make the usual distinction between empirical laws and scientific theories, we could say that the kind of explanation I have discussed is explanation by means of empirical laws. For all that has been said in this paper, theoretical explanation—explanation that makes use of scientific theories in the fullest sense of the term—may have a logical structure entirely different from that of statistical explanation. Although theoretical explanation is almost certainly the most important kind of scientific explanation, it does, nevertheless, seem useful to have a clear account of explanation by means of empirical laws, if only as a point of departure for a treatment of theoretical explanation.

Second, in remarking above that statistical explanation is nomological, I was tacitly admitting that the statistical or universal generalizations invoked in explanations should be lawlike. I have made no attempt to analyze lawlikeness, but it seems likely that an adequate analysis will involve a solution to Nelson Goodman's "grue-bleen" problem.[79]

Third, my account of statistical explanation obviously depends heavily upon the concept of *statistical relevance* and upon the *screening-off relation,* which is defined in terms of statistical relevance. In the course of the discussion, I have attempted to show how these tools enable us to capture much of the involvement of explanation with causality, but I have not attempted to provide an analysis of causation in terms of these statistical concepts alone. Reichenbach has attempted such an analysis,[80] but whether his—or any other—can succeed is a difficult question. I should be inclined to harbor serious misgivings about the adequacy of my view of statistical explanation if the statistical analysis of causation cannot be carried through successfully, for the relation between causation and explanation seems extremely intimate.

Finally, although I have presented my arguments in terms of the limiting frequency conception of probability, I do not believe that the fundamental correctness of the treatment of statistical explanation hinges upon the acceptability of that interpretation of probability. Proponents of other theories of probability, especially the personalist and the propensity interpretations, should be able to adapt this treatment of explanation to their views of probability with a minimum of effort. That, too, is left as an exercise for the reader.[81]

NOTES

1. Carl G. Hempel and Paul Oppenheim, "Studies in the Logic of Explanation," *Philosophy of Science,* XV (1948), pp. 135–75. Reprinted, with a 1964 "Postscript," in Carl G. Hempel, *Aspects of Scientific Explanation* (New York: Free Press, 1965).
2. Carl G. Hempel, "Explanation in Science and in History," in *Frontiers in Science and Philosophy,* ed. Robert G. Colodny (Pittsburgh: University of Pittsburgh Press, 1962), p. 10.
3. See also Hempel, *Aspects of Scientific Explanation,* pp. 367–68, where he offers "a general *condition of adequacy for any rationally acceptable explanation of a particular event,*" namely, that "any rationally acceptable answer to the question 'why did event *X* occur?' must offer information which shows that *X* was to be expected—if not definitely, as in the case of D-N explantion, then at least with reasonable probability."

   Inductive explanations have variously been known as "statistical," "probabilistic," and "inductive-statistical." Deductive explanations have often been called "deductive-nomological." For the present I shall simply use the terms "inductive" and "deductive" to emphasize the crucial fact that the former embody inductive logical relations, whereas the latter embody deductive logical relations. Both types are nomological, for both require lawlike generalizations among their premises. Later on, I shall use the term "statistical explanation" to refer to the sort of explanation I am trying to characterize, for it is statistical in a straightforward sense, and it is noninductive in an extremely important sense.
4. Rudolf Carnap, "Preface to the Second Edition," in *Logical Foundations of Probability* (Chicago: University of Chicago Press, 1962), 2d ed., pp. xv–xx
5. *Aspects of Scientific Explanation.*
6. "Deductive-Nomological vs. Statistical Explanation," in *Minnesota Studies in the Philosophy of Science,* III, eds. Herbert Feigl and Grover Maxwell (Minneapolis: University of Minnesota Press, 1962).
7. I called attention to this danger in "The Status of Prior Probabilities in Statistical Explanation," *Philosophy of Science,* XXXII, no. 2 (April, 1965), p. 137. Several fundamental disanalogies could be cited. First, the relation of deductive entailment is transitive, whereas the relation of inductive support is not; see my "Consistency, Transitivity, and Inductive Support," *Ratio,* VII, no. 2 (Dec. 1965), pp. 164–69. Second, on Carnap's theory of degree of confirmation, which is very close to the notion of inductive probability that Hempel uses in characterizing statistical explanation, there is no such thing as inductive inference in the sense of allowing the detachment of inductive conclusions in a manner analogous

to that in which deductive logic allows the detachment of conclusions of deductive inferences. See my contribution "Who Needs Inductive Acceptance Rules?" to the discussion of Henry E. Kyburg's "The Rule of Detachment in Inductive Logic," in *The Problem of Inductive Logic*, ed. Imre Lakatos (Amsterdam: North Holland Publishing Co., 1968), pp. 139–44, for an assessment of the bearing of this disanalogy specifically upon the problem of scientific explanation. Third, if q follows from p by a deductively valid argument, then q follows validly from p & r, regardless of what statement r is. This is the reason that the *requirement of total evidence* is automatically satisfied for deductive-nomological explanations. By contrast, even if p provides strong inductive support for q, q may not be inductively supported at all by p & r. Informally, a valid deductive argument remains valid no matter what premises are added (as long as none is taken away), but addition of premises to a strong inductive argument can destroy all of its strength. It is for this reason that the *requirement of total evidence* is not vacuous for statistical explanations.

8. See "Deductive-Nomological vs. Statistical Explanation."
9. See, for example, Hempel, "Explanation in Science and in History."
10. In the present context nothing important hinges upon the particular characterization of the parts of arguments. I shall refer to them indifferently as statements or propositions. Propositions may be regarded as classes of statements; so long as they are not regarded as facts of the world, or nonlinguistic states of affairs, no trouble should arise.
11. When no confusion is apt to occur, we may ignore the distinction between the explanandum and the explanandum event. It is essential to realize, however, that a given explanation must not purport to explain the explanandum event in all of its richness and full particularity; rather, it explains just those aspects of the explanandum event that are mentioned in the explanandum.
12. "Deductive-Nomological vs. Statistical Explanation," p. 125.
13. Ibid. See also "Inductive Inconsistencies," *Synthèse*, XII, no. 4 (Dec. 1960).
14. Hempel and Oppenheim, "Studies in the Logic of Explanation," and Hempel, "Deductive-Nomological vs. Statistical Explanation."
15. This condition has sometimes been weakened to the requirement that the premises be highly confirmed. I prefer the stronger requirement, but nothing very important hangs on the choice. See n. 76.
16. A premise occurs essentially in an argument if that argument would cease to be (deductively or inductively) correct upon deletion of that premise. Essential occurrence does not mean that the argument could not be made logically correct again by replacing the premise in question with another premise. "Essential occurrence" means that the premise plays a part in the argument as given; it does not just stand there contributing nothing to the logical correctness of the argument.
17. The requirement of total evidence demands that there should be no additional statements among our available stock of statements of evidence that would change the degree to which the conclusion is supported by the argument if they were added to the argument as premises. See Carnap, *Logical Foundations of Probability*, sec. 45B.
18. Hempel, "Explanation in Science and in History," p. 14.
19. "The Status of Prior Probabilities in Statistical Explanation," p. 145.
20. Consumer Reports, *The Medicine Show* (New York: Simon & Schuster, 1961), pp. 17–18 (*Pace* Dr. Linus Pauling).
21. Henry E. Kyburg, "Comments," *Philosophy of Science*, XXXII, no. 2 (April 1965), pp. 147–51.

22. Since there is no widely accepted account of inductive inference, it is difficult to say what constitutes correct logical form for inductive arguments. For purposes of the present discussion I am accepting Hempel's schema (4) as one correct inductive form.
23. The fact that the general premise in this argument refers explicitly to a particular physical object, the moon, may render this premise nonlawlike, but presumably the explanation could be reconstructed with a suitably general premise about satellites.
24. In Hempel and Oppenheim's "Studies in the Logic of Explanation," (7.5) constitutes a set of necessary but not sufficient conditions for a potential explanans, while (7.6) defines *explanans* in terms of potential explanans. However, (7.8) provides a definition of *potential explanans,* which in combination with (7.6) constitutes a set of necessary and sufficient conditions for an explanans.
25. *Aspects of Scientific Explanation,* pp. 397–403. See also Carl G. Hempel, "Maximal Specificity and Lawlikeness in Probabilistic Explanation," *Philosophy of Science,* XXXV (1968), pp. 116–33, which contains a revision of this requirement. The revision does not seem to affect the objections I shall raise.
26. Salmon, "The Status of Prior Probabilities."
27. This characterization of the logical interpretation of probability is patterned closely upon Carnap, *Logical Foundations of Probability.*
28. Salmon, "Who Needs Inductive Acceptance Rules?".
29. This point was brought out explicitly and forcefully by Richard Jeffrey in "Statistical Explanation vs. Statistical Inference" presented at the meeting of the American Association for the Advancement of Science, Section L, New York, 1967. This superb paper has since been published in *Essays in Honor of Carl G. Hempel,* ed. Nicholas Rescher (Dordrecht, Holland: Reidel Publishing Co., 1969), and it is reprinted herein. Hempel discussed this issue in "Deductive-Nomological vs. Statistical Explanation," pp. 156–63.
30. For an indication of some of the difficulties involved in properly circumscribing this body of total evidence, see Kyburg's comments on my paper "The Status of Prior Probabilities in Statistical Explanation," and my rejoinder. Hempel has discussed this problem at some length in "Deductive-Nomological vs. Statistical Explanation," pp. 145–49, and "Aspects of Scientific Explanation," sec. 3.4.
31. An extremely clear account of this view is given by Ward Edwards, Harold Lindman, and Leonard J. Savage in "Bayesian Statistical Inference for Psychological Research," *Psychological Review,* LXX, no. 3 (May 1963). An approach which is very similar in some ways, though it is certainly not a genuinely subjective interpretation in the Bayesian sense, is given by Carnap in "The Aim of Inductive Logic," in *Logic, Methodology and Philosophy of Science,* eds. Ernest Nagel, Patrick Suppes, and Alfred Tarski (Stanford, Calif.: Stanford University Press, 1962), pp. 303–18.
32. This treatment of the frequency interpretation takes its departure from Hans Reichenbach, *The Theory of Probability* (Berkeley and Los Angeles: University of California Press, 1949), but the discussion differs from Reichenbach's in several important respects, especially concerning the problem of the single case. My view of this matter is detailed in *The Foundations of Scientific Inference* (Pittsburgh: University of Pittsburgh Press, 1967), pp. 83–96.
33. See *Foundations of Scientific Inference,* pp. 56–96, for discussions of the various interpretations of probability.
34. See, for example, Karl Popper, "The Propensity Interpretation of the Calculus of Probability, and the Quantum Theory," in *Observation and Interpretation,* ed. S. Körner (London: Butterworth Scientific Publications, 1956), pp. 65–70, and

"The Propensity Interpretation of Probability," *British Journal for the Philosophy of Science,* X (1960), pp. 25–42.

35. See Salmon, *The Foundations of Scientific Inference,* pp. 83–96, for fuller explanations. Note that, contrary to frequent usage, the expression "$P(A,B)$" is read "the probability *from A to B.*" This notation is Reichenbach's.
36. Reichenbach, *The Theory of Probability,* sec. 72. John Venn, *The Logic of Chance,* 4th ed. (New York: Chelsea Publishing Co., 1962), chap. IX, sec. 12–32. Venn was the first systematic exponent of the frequency interpretation, and he was fully aware of the problem of the single case. He provides an illuminating account, and his discussion is an excellent supplement to Reichenbach's well-known later treatment.
37. Reichenbach, *The Theory of Probability,* p. 374.
38. Carnap, *Logical Foundations of Probability,* sec. 44.
39. Richard von Mises, *Probability, Statistics and Truth,* 2d rev. ed. (London: Allen and Unwin, 1957), p. 25.
40. Also, of course, there are cases in which it would be possible in principle to make a relevant partition, but we are playing a game in which the rules prevent it. Such is the case in roulette, where the croupier prohibits additional bets after a certain point in the spin of the wheel. In these cases also we shall speak of practical homogeneity.
41. "Deductive-Nomological vs. Statistical Explanation," p. 133. Hempel also presents a résumé of these considerations in *Aspects of Scientific Explanation,* pp. 394–97.
42. "Deductive-Nomological vs. Statistical Explanation," pp. 146–47.
43. Ibid.
44. *Aspects of Scientific Explanation,* pp. 399–400. Note that in Hempel's notation "$p(G,F,)$" denotes "the probability of G, given F." This is the reverse of Reichenbach's notation.
45. Ibid., pp. 400–01. In a footnote to this passage Hempel explains why he prefers this requirement to the earlier requirement of total evidence.
46. In working out the technical details of the following definition, it is necessary to recognize that homogeneous partitions are not unique. The transfer of a finite number of elements from one compartment to another will make no difference in any probability. Similarly, even an infinite subclass C_j, if $P(A,C_j) = 0$, can fulfill the condition that $P(A.C_j,B) = p_j \neq P(A,B)$ without rendering A's degree of inhomogeneity > 0. Such problems are inherent in the frequency approach, and I shall not try to deal with them here. In technical treatment one might identify subclasses that differ by measure zero, and in some approaches the conditional probability $P(A.C_j,B)$ is not defined if $P(A,C_j) = 0$.
47. I am grateful to Prof. Douglas Stewart of the Department of Sociology at the University of Pittsburgh for pointing out that the concept of variance is of key importance in this context.
48. See James G. Greeno, "Evaluation of Statistical Hypotheses Using Information Transmitted," *Philosophy of Science,* XXXVII (1970), reprinted herein as "Explanation and Information," for a precise and quantitative discussion of this point.
49. See Michael J. Scriven, "Explanation and Prediction in Evolutionary Theory," *Science,* CXXX, no. 3374 (Aug. 28, 1959).
50. Ibid., p. 480.
51. See Adolf Grünbaum, *Philosophical Problems of Space and Time* (New York: Alfred A. Knopf, 1963), pp. 309–11.
52. Hans Reichenbach, *The Direction of Time* (Berkeley and Los Angeles: The University of California Press, 1956) p. 189.

53. Since $P(A.C.B) = P(A,B)$ entails $P(A.B,C) = P(A,C)$, provided $P(A.B,C) \neq 0$, the relevance relation is symmetrical. The screening-off relation is a three-place relation; it is nonsymmetrical in its first and second arguments, but it is symmetrical in the second and third arguments. If D screens off C from B, then D screens off B from C.
54. Reichenbach, *The Direction of Time*, sec. 22.
55. This example has received considerable attention in the recent literature on explanation. Introduced in Scriven, "Explanation and Prediction," it has been discussed by (among others) May Brodbeck, "Explanation, Prediction, and 'Imperfect Knowledge,'" in *Minnesota Studies in Philosophy of Science*, III, eds. Herbert Feigl and Grover Maxwell (Minneapolis: University of Minnesota Press, 1962); Adolf Grünbaum, *Philosophical Problems*, pp. 303–08; and Carl G. Hempel, "Explanation and Prediction by Covering Laws," *Philosophy of Science: The Delaware Seminar*, I, ed. Bernard H. Baumrin (New York: John Wiley and Sons, 1963).
56. "72 out of 100 untreated persons [with latent syphilis] go through life without the symptoms of late [tertiary] syphilis, but 28 out of 100 untreated persons were known to have developed serious outcomes [paresis and others] and there is no way to predict what will happen to an untreated infected person" (Edwin Gurney Clark, M.D., and William D. Mortimer Harris, M.D., "Venereal Diseases," *Encyclopedia Britannica*, XXIII [1961], p. 44).
57. Ralph E. Lapp and Howard L. Andrews, *Nuclear Radiation Physics*, 3rd ed. (Englewood Cliffs, N.J.: Prentice-Hall, 1963), p. 73.
58. Cf. Hempel's discussion of this example in "Aspects of Scientific Explanation," pp. 369–74.
59. Here I am assuming the separate trials to be independent events.
60. George Gamow, *The Atom and Its Nucleus* (Englewood, Cliffs, N.J.: Prentice-Hall, 1961), p. 114.
61. Cf. Hempel's discussion in "Aspects of Scientific Explanation," sec. 4.2.2, of an example of a slip of the pen taken from Freud. Hempel takes such examples to be partial explanations.
62. See Jeffrey, "Statistical Explanation vs. Statistical Inference" for a lucid and eloquent discussion of this point. In this context I am, of course, assuming that the pertinent probabilities exist.
63. Lapp and Andrews, *Nuclear Radiation Physics*, p. 73.
64. I owe an enormous intellectual debt to Reichenbach and Grünbaum in connection with the view of explanation offered in this paper. Such examples as the paradigm cases of explanation to be discussed have been subjected to careful analysis by these men. See Grünbaum, *Philosophical Problems*, chap. IX, and Reichenbach, *The Direction of Time*, chaps. III–IV, for penetrating and lucid analyses of these examples and the issues raised by them.
65. In his *Theory of Probability*, sec. 34, Reichenbach introduces what he calls the *probability lattice* as a means for dealing with sequences in which we want to say that the probabilities vary from element to element. One particular type of lattice, the *lattice of mixture* is used to describe the temporally asymmetric character of the ensemble of branch systems. The *time ensemble* and the *space ensemble* are characterized in terms of the lattice arrangement. See Reichenbach, *The Direction of Time*, sec. 14.
66. Ibid., sec. 18.
67. Ibid., chap. IV.
68. See Hempel, "Deductive-Nomological vs. Statistical Explanation," pp. 109–10; also Grünbaum, *Philosophical Problems*, pp. 307–08.

69. Reichenbach, *The Direction of Time,* sec. 6.
70. Carnap, *Logical Foundations of Probability,* sec. 44.
71. Ibid., secs. 50–51.
72. Salmon, "Who Needs Inductive Acceptance Rules?" Because of this difference with Carnap—i.e., my claim that inductive logic requires rules of acceptance for the purpose of establishing statistical generalizations—I do not have the thoroughgoing "pragmatic" or "instrumentalist" view of science Hempel attributes to Richard Jeffrey and associates with Carnap's general conception of inductive logic. Cf. Hempel, "Deductive-Nomological vs. Statistical Explanation," pp. 156–63.
73. Salmon, *Foundations of Scientific Inference,* pp. 90–95.
74. See "Aspects of Scientific Explanation," secs. 3.2–3.3. In the present essay I am not at all concerned with explanations of the type Hempel calls "deductive-statistical." For greater specificity, what I am calling "statistical explanation" might be called "statistical-relevance explanation," or "S-R explanation" as a handy abbreviation to distinguish it from Hempel's D-N, D-S, and I-S types.
75. Hempel, "Deductive-Nomological vs. Statistical Explanation," sec. 3.
76. Rudolf Carnap, "Truth and Confirmation," in *Readings in Philosophical Analysis,* eds. Herbert Feigl and Wilfrid Sellars (New York: Appleton-Century-Crofts, 1949), pp. 119–27.
77. Reichenbach, *Theory of Probability,* sec. 69.
78. Hempel, "Deductive-Nomological vs. Statistical Explanation," sec. 13, and "Aspects of Scientific Explanation," sec. 3.6. Here, Hempel says, "Nonconjunctiveness presents itself as an inevitable aspect of [inductive-statistical explanation], and thus as one of the fundamental characteristics that set I-S explanation apart from its deductive counterparts."
79. See Nelson Goodman, *Fact, Fiction, and Forecast,* 2d ed. (Indianapolis: Bobbs-Merrill Co., 1965), chap. III. I have suggested a resolution in "On Vindicating Induction," *Philosophy of Science,* XXX (July 1963), pp. 252–61, reprinted in Henry E. Kyburg and Ernest Nagel, eds., *Induction: Some Current Issues* (Middletown, Conn.: Wesleyan University Press, 1963).
80. Reichenbach, *The Direction of Time,* chap. IV.
81. The hints are provided in sec. 3.

Explanation and Information

JAMES G. GREENO

University of Michigan

The main argument of this paper is that an evaluation of the overall explanatory power of a theory is less problematic and more relevant as an assessment of the state of knowledge than evaluation of statistical explanations of single occurrences in terms of likelihoods that are assigned to explananda.

One purpose of this paper is to suggest a way around an apparent paradox in the current theory of statistical explanation. The paradox has been mentioned in earlier analyses [2] and can be illustrated by the following situation. Suppose that a boy, Albert, is convicted for stealing a car. Attempting to give an explanation, a social worker points out that Albert lives in San Francisco, where there is a high delinquency rate. However, it is also noted that Albert's father earns $40,000 per year, and sons of men with high incomes have a low delinquency rate.

One view which agrees with intuition and most recent discussions of statistical explanation, is that a "good" statistical explanation allows us to assign a high probability or likelihood to the event to be explained. Let S be satisfied by a young man living in San Francisco, and let M be satisfied if a young man is convicted of a major crime. Then we have the statistical hypothesis

$$P(M \mid S) = p,$$

where p is some rather high probability. And we can carry out a simple argument of the form

$$\text{(1)} \qquad \frac{P(M \mid S) = p, \quad a \in S,}{L(a \in M) = p},$$

where $L(a \in M)$ stands for the likelihood that individual a is in the set (has the property) M. Since we take it that p is fairly high, argument (1) should be taken as a fairly good statistical explanation of Albert's delinquency.

The difficulty with this is that we have not included an item of relevant information, the income-level of Albert's father. Let H be satisfied if a boy comes from a family with more than, say, $30,000 per year. Then we have

$$P(M \mid S \cap H) = p',$$

where p' is considerably lower than p (and probably lower than $P(M)$ in the whole population). And the argument becomes

$$\text{(2)} \qquad \frac{P(M \mid S \cap H) = p' \quad a \in S \cap H,}{L(a \in M) = p'}.$$

* Received August, 1968.

According to Salmon [4] argument (2) is closer to the correct explanation of Albert's delinquency than argument (1), because (1) does not assign Albert to a homogeneous class with respect to the event to be explained. Salmon has argued that the best statistical explanation is one that assigns the correct likelihood to the explanandum, and Salmon's point certainly rings true. However, it seems counter-intuitive to assert that an argument like (2) is a better explanation of $a \in M$ than an argument like (1), since (2) leads to the conclusion that the explanandum was unlikely.

I do not see any way of resolving this paradox as long as we try to evaluate our state of knowledge by looking at single statistical explanations. However, there do not seem to be any compelling reasons for doing that. Salmon's comment suggests that any single statistical explanation should not be considered as "good" or "bad" because of the value of the likelihood assigned to the explanandum. The relevant question about a single statistical explanation is just whether the likelihood assigned to the explanandum is correct or incorrect. This question may be empirical, regarding the accuracy of hypotheses, or formal, regarding the validity of arguments. But these questions about single explanations do not seem to involve fundamental philosophical issues.

In the view to be developed here, questions that are relevant to the methodology of science arise with regard to general explanatory systems, rather than single explanations. I will discuss the problem of evaluating states of knowledge in relation to theories that include statistical hypotheses. Then the paradox of evaluation in terms of likelihoods is avoided, because we no longer evaluate in terms of single likelihoods. The burden of the argument will be to show that the alternative of evaluating general systems agrees with intuition in some important ways and leads to some clarification of the nature of scientific knowledge.

1. General Considerations. For the purposes of this paper, a theory is defined as a set of assumptions that specify a domain, a list of descriptive predicates and a probability measure on the events of the domain. Formally, a theory is a triple $\langle \Omega, \mathscr{A}, P \rangle$ where Ω is the domain, $\mathscr{A}$ is a Borel field of the subsets of Ω, and P is a probability function with domain $\mathscr{A}$. For example, in a theory about juvenile deliquency, the domain might be American males between 12 and 18 years old. The variables of the theory might be degrees of delinquency D, the kind of neighborhood a boy lives in B, and the income of the boy's family C. Let the values of D be

D_1 = no convictions,
D_2 = minor convictions only,
D_3 = major convictions.

Let the values of B be

B_1 = rural or small town,
B_2 = small city,
B_3 = suburb of large city,
B_4 = center city of a metropolitan area.

And let the values of C be

C_1 = less than \$4000,
C_2 = \$4000–\$10,000,
C_3 = \$10,000–\$30,000,
C_3 = over \$30,000.

The class of subsets $\mathscr{A}$ is the product $\{D\} \times \{B\} \times \{C\}$, and p assigns a probability to each event that is a member of $\mathscr{A}$. For example, from the assumptions of a theory, we could derive $P(D_1 \cap B_3 \cap C_3)$, the probability that an American male between 12 and 18 years old has no convictions, lives in a suburb, and comes from a family with an income between \$10,000 and \$30,000.

I am mainly interested in suggesting a rough metric for evaluating a theory of the kind I have described. The metric is an elementary quantity in statistical information theory, the information transmitted by the system. As is well known, the statistical concept of information is based on the unpredictability or uncertainty of events. The amount of information transmitted is the amount by which uncertainty is reduced because of dependencies among events. Therefore, the notion of information transmitted is a natural candidate for evaluating the explanatory power of a theory about the events.

To apply the information-theoretic ideas, I will consider a theory in terms of two sets of variables $\{S\}$ and $\{M\}$. $\{M\}$ is an m-cell partition of Ω, and the intended interpretation is a variable or a set of variables whose values are to be explained. $\{M\}$ will be called the explananda of the theory. $\{S\}$ is an n-cell partition of Ω, and the intended interpretation is a set of variables used to explain occurrences of the values of $\{M\}$. $\{S\}$ will be called the explanans of the theory. In the example about deliquency, we ordinarily would use $\{D\} = \{D_1, \ldots, D_3\}$ as the explananda, and $\{B\} \times \{C\} = \{B_1C_1, \ldots, B_4C_4\}$ as the explanans. For the general discussion here, I will denote the explanans and the explananda as

$$\{S\} = \{S_1, \ldots, S_i, \ldots, S_n\}, \qquad \{M\} = \{M_1, \ldots, M_j, \ldots, M_m\}.$$

The field $\mathscr{A}$ is clearly $\{S\} \times \{M\}$.

The information transmitted in this kind of system is

$$I_T = H(S) + H(M) - H(S \times M), \tag{3}$$

$H(S)$ is the uncertainty of the explanans,

$$H(S) = \sum_{i=1}^{n} -p_i \log p_i,$$

where p_i is the probability of S_i. $H(M)$ is the uncertainty of the explananda,

$$H(M) = \sum_{j=1}^{m} -p_j \log p_j,$$

where p_j is the probability of M_j. And $H(S \times M)$ is the uncertainty of the system,

$$H(S \times M) = \sum_{i=1}^{n} \sum_{j=1}^{m} -p_i p_{ij} \log p_i p_{ij},$$

where p_{ij} is the conditional probability of M_j given S_i.

Two remarks can be made regarding some general properties of information transmitted as an index of explanatory power. The first remark shows that the value of I_T is closely related to the extent to which the theory identifies variables that are relevant to the understanding of the explananda. The second remark deals with a situation where the theory permits explanations and predictions in a trivial way: the value of I_T is low in a situation where this happens. These two remarks contribute to the argument that I_T provides an evaluation of a theory that agrees with intuition.

Relevance of Variables—A theory applies to some domain that is partitioned by the explananda—for example, the population of the United States is divided into Republicans, Democrats, and Independents in some proportions. A theory that attempts to explain the values of M identifies variables that are relevant to values of M. We would say that a variable S was irrelevant to M if S and M were independent. S is a relevant variable for explaining M if the conditional probabilities of M_j given S_i are not equal to the marginal probabilities of M_j. For example, the section of the country where a person lives is relevant to his political party affiliation—people who live in the South tend to be Democrats and people who live in the Southwest tend to be Republicans, more than the general population. It seems likely that a person's height is not relevant to his party membership. We would be surprised if the proportions of Democrats, Republicans, and Independents among tall people were different from the proportions among the population as a whole.

The value of I_T is non-negative, and tends to be large if the explanans and explananda are strongly dependent. The value of I_T is zero if the explanans and explananda are independent. Independence implies that $p_{ij} = p_j$ for all i. Then

$$
\begin{aligned}
H(S \times M) &= \sum_{i=1}^{n} \sum_{j=1}^{m} -p_i p_j \log p_i p_j \\
&= \sum_{i=1}^{n} \sum_{j=1}^{m} -p_i p_j (\log p_i + \log p_j) \\
&= \sum_{i=1}^{n} -p_i \log p_i \sum_{j=1}^{m} p_j + \sum_{j=1}^{m} -p_j \log p_j \sum_{i=1}^{n} p_i \\
&= H(S) + H(M),
\end{aligned}
$$

which makes $I_T = 0$.

The value of I_T is maximal if all the p_{ij} are either one or zero. Consider the following example.

$$
(4) \qquad P(S_i \cap M_j) = \begin{array}{ccc} & M_1 & M_2 \\ S_1 & 0.50 & 0.00 \\ S_2 & 0.00 & 0.50. \end{array}
$$

Using natural logarithms,

$$H(S) = H(M) = 2(-0.50) \log (0.50) = 0.69.$$

But

$$H(S \times M) = (2)[(-0.50) \log (0.50) + (0.0) \log (0.0)].$$

It can be shown that

$$\lim_{p \to 0} p \log p = 0.$$

Therefore, whenever the values of p_{ij} go to one and zero, $H(S \times M)$ goes to $H(S)$, and the value of I_T goes to its maximum, $H(M)$.

These results agree with intuition. A theory with $I_T = 0$ does not allow us to select cases with different values of M any better than we could just by knowing the probabilities of the values of M. On the other hand, the situation that we would like is one that allows nomological-deductive prediction, and in that case all of the values of p_{ij} are either one or zero, which makes I_T take its maximum value. In general, I_T is somewhere between zero and $H(M)$, and the value of I_T gives an indication of the extent to which the theory is "close" to the goal of nomological deductive explanatory power.

Dependence of I_T on $H(M)$ and $H(S)$—It is possible to construct imaginary theories that have all values of p_{ij} equal to one or zero, but which nonetheless lack explanatory power. Theories can be trivial for many reasons, but one way to produce triviality would be to try to explain a characteristic that is a property of all members of the domain. For example, physical anthropologists are not interested in explaining why the members of some primitive tribes have two legs—all humans have two legs. Suppose a theory specified the following probability measure:

$$P(S_i \cap M_j) = \begin{array}{c|cc} & M_1 & M_2 \\ S_1 & 0.50 & 0.00 \\ S_2 & 0.50 & 0.00 \end{array} \qquad (5)$$

In both equations (4) and (5) all the values of p_{ij} are either one or zero. However, in equation (4) the explanans really explain things. One-half of the members of the domain have the property M_1, and the explanans allow us to select just those cases. But in equation (5), all of the members of the domain have the property M_1, so the explanans are not doing any work.

The point can be developed more exactly using an alternative form of equation (3). The conditional uncertainty of M given S_i is defined as

$$H(M \mid S_i) = \sum_{j=1}^{m} -p_{ij} \log p_{ij}.$$

This measures the unpredictability of the explananda for the subset of the domain where $S = S_i$. Then we can show that

$$I_T = H(M) - \sum_{i=1}^{n} p_i H(M \mid S_i). \qquad (6)$$

The result can be proved as follows:

$$H(S \times M) = \sum_{i=1}^{n} \sum_{j=1}^{m} -p_i p_{ij}(\log p_i + \log p_{ij})$$

$$= \sum_{i=1}^{n} -p_i \log p_i \sum_{j=1}^{m} p_{ij} + \sum_{i=1}^{n} p_i \sum_{j=1}^{m} -p_{ij} \log p_{ij}$$

$$= H(S) + \sum_{i=1}^{n} p_i H(M \mid S_i),$$

and then the result follows directly from equation (3).

Equation (6) says that the information transmitted cannot be greater than the uncertainty of the explananda. In effect, if we can do quite a good job explaining the explananda without the theory, then the theory cannot help as much as it could if the explananda were more uncertain. For example, it could be quite interesting and useful to investigate factors that relate to whether a person from a middle-class family attends college or not—some do and others do not. On the other hand, there would be less opportunity to increase the information transmitted by investigating factors that might relate to college attendance by sons of upper-class families, since almost all of these young men attend college. Of course, this does not imply that it will always, or even usually, be unprofitable to investigate hypotheses about rare events. Another remark about this issue will be made in section **3**.

A related fact involves a symmetry in explanatory systems. Equation (6) expresses the information transmitted in terms of the uncertainty of the explananda and their conditional uncertainties given the explanans. By a similar argument, it can be shown that

$$(7) \qquad I_T = H(S) - \sum_{j=1}^{m} p_j H(S \mid M_j)$$

where

$$H(S \mid M_j) = \sum_{i=1}^{n} -p_{ji} \log p_{ji}, \qquad p_{ji} = P(S_i \mid M_j).$$

Equation (6) relates to the fact that it is trivial to explain characteristics of the entire domain. Equation (7) relates to the fact that it is impossible to obtain any explanatory power using an explanans that has only one value in the population. For example, since all humans are bipeds, the number of legs that people have cannot be used to explain any differences among people (Though of course it can be used to explain some differences between humans and, say, horses.)

To summarize these remarks, it appears that evaluating an explanatory system by using information transmitted agrees with intuition in that stronger dependencies increase the value of I_T. There are limits on the value of I_T that are relatively restrictive when the probabilities of either the explananda or the explanans are very high or very low, and these limits make sense in relation to the triviality of explaining universal characteristics and the uselessness of universal characteristics as explanans.

2. Relationship to Predictive Usefulness. This section deals with the relationship between information transmitted and the usefulness of the theory in permitting accurate predictions. To develop the point, begin with a simple illustration. Suppose that M is binary, so that for each value of S there are two values of p_{ij}. Given a value of S we have the probabilities of M_1 and M_2, namely p_{i1} and p_{i2}. Now consider the following rule for making predictions. Predict M_1 in a proportion of cases equal to p_{i1}, and predict M_2 in the remaining cases.

A general version of the rule mentioned above is the following: For all M_j $(j = 1, \ldots, m)$ predict M_j in a proportion of cases equal to p_{ij} when the value of S is S_i. This does not lead to the maximum number of correct predictions. The optimal rule is to find the maximum value of p_{ij} and predict M_j for all cases where $S = S_i$. However, analysis of the optimal rule requires strong assumptions about all the values of p_{ij} and the resulting complexity would make the whole situation much cloudier.

Under the proposed rule for making predictions, the proportion of correct predictions when $S = S_i$ will equal

$$P(\text{correct} \mid S_i) = \sum_{j=1}^{m} p_{ij}^2,$$

and the overall proportion of correct predictions will equal

$$c_1 = P(\text{correct} \mid S) = \sum_{i=1}^{n} p_i \sum_{j=1}^{m} p_{ij}^2.$$

This quantity can be compared with the proportion of correct predictions that would be made using the same rule, but without using the explanans. Let

$$c_0 = \sum_{j=1}^{m} p_j^2.$$

Then the improvement in overall predictive accuracy due to the theory is

$$c_1 - c_0 = \sum_{i=1}^{n} p_i \sum_{j=1}^{m} p_{ij}^2 - \sum_{j=1}^{m} p_j^2. \tag{8}$$

Equation (8) can be compared with the amount of information transmitted by the theory, in the form given by equation (6).

The Taylor expansion of $-x \log x$ is

$$-x \log x = x(1 - x) + \frac{x(1 - x)^2}{2} + \frac{x(1 - x)^3}{3} + \cdots.$$

In equation (6), we have

$$H(M) = \sum_{j=1}^{m} -p_j \log p_j = \sum_{j=1}^{m} p_j(1 - p_j) + R_2(p_j)$$

where

$$0 < R_2(p_j) < \tfrac{1}{2} \sum_{j=1}^{m} (1 - p_j)^2.$$

The other term in equation (6) is

$$\sum_{i=1}^{n} p_i H(M \mid S_i) = \sum_{i=1}^{n} p_i \sum_{j=1}^{m} -p_{ij} \log p_{ij} = \sum_{i=1}^{n} p_i \sum_{j=1}^{m} p_{ij}(1 - p_{ij}) + R_2(p_{ij}^2)$$

where

$$0 < R_2(p_{ij}) < \tfrac{1}{2} \sum_{i=1}^{n} p_i \sum_{j=1}^{m} (1 - p_{ij})^2.$$

Combining the terms, we obtain

$$I_T = \sum_{j=1}^{m} p_j(1 - p_j) - \sum_{i=1}^{n} p_i \sum_{j=1}^{m} p_{ij}(1 - p_{ij}) + [R_2(p_j) - R_2(p_{ij})]$$

$$= \left[\sum_{i=1}^{n} p_i \sum_{j=1}^{m} p_{ij}^2 - \sum_{j=1}^{m} p_j^2\right] + [R_2(p_j) - R_2(p_{ij})]$$

(10) $$I_T = (c_1 - c_0) + [R_2(p_j) - R_2(p_{ij})].$$

The result shows that, to a reasonable approximation, the information transmitted is equal to the usefulness of the theory in improving the accuracy of predictions about the explananda.

To illustrate the ideas and give some idea about the approximation, consider the following examples:

(11) $P(S_i \cap M_j) =$

	M_1	M_2
S_1	0.30	0.20
S_2	0.20	0.30

(12) $P(S_i \cap M_j) =$

	M_1	M_2
S_1	0.40	0.10
S_2	0.10	0.40

In both equations, we have

$$c_0 = 0.25 + 0.25 = 0.50.$$

In equation (11), we have

$$c_1 = 0.36 + 0.16 = 0.52.$$

$$c_1 - c_0 = 0.0200.$$

The information transmitted is

$$I_T = 0.0202.$$

In equation (12), we have

$$c_1 = 0.64 + 0.04 = 0.68,$$

$$c_1 - c_0 = 0.180.$$

The information transmitted is

$$I_T = 0.193.$$

When we analyze the overall predictive usefulness of a theory using a rule like the one assumed here, we suppose that correct predictions are equally important regardless of what value of M occurs. This seems reasonable from the point of view of theoretical science, but it surely is not correct in many practical situations. For example, if a medical theory is developed to explain why a certain disease occurs, the theory's predictive usefulness probably depends much more on its usefulness in predicting which people will contract the disease than on its usefulness in predicting which people will remain healthy.

The point can be illustrated by the following example:

$$P(S_i \cap M_j) = \begin{array}{ccc} & M_1 & M_2 \\ S_1 & 0.990 & 0.000 \\ S_2 & 0.009 & 0.001 \end{array} \tag{13}$$

S_2 might be the event of being exposed to some communicable disease and M_2 the event of contracting the disease. Scriven [5] has used similar examples in criticizing the symmetry between prediction and explanation implied by Hempel's [2] analysis. The theory expressed by (13) does not permit very reliable positive predictions about any individual's contracting the disease, but it does tell us something quite important about the disease. Also, as Salmon [4] has pointed out, the theory does permit a large number of reliable negative predictions.

Another example, involving the same value of I_T as equation (13) is the following:

$$P(S_i \cap M_j) = \begin{array}{ccc} & M_1 & M_2 \\ S_1 & 0.000 & 0.990 \\ S_2 & 0.001 & 0.009 \end{array} \tag{14}$$

In equation (14) M_2 might be the event of favoring a war in which one's government is engaged, and S_2 might be the event of being exposed to speeches made by people who opposed the war on moral grounds. In this case, the theory can be used to make many reliable positive predictions about individuals who favor the war, but not many reliable negative predictions. The fact that equations (13) and (14) give the same value of I_T agrees with the intuitive judgment that the two cases involve equal explanatory power about the events involved. It is clear that the story is incomplete in both cases. Further knowledge would be needed to distinguish the cases of S_1 in which M_1 occurs from those in which M_2 occurs.

But the main point of these examples is that a theory with a nonzero value of I_T may or may not permit very reliable predictions about a single explanandum that we happen to be interested in for practical reasons. The relationship between prediction and explanation seems to be clarified somewhat by emphasizing the fact that predictions and explanations of single occurrences involve both positive and negative cases, while the explanatory power of the theory can be evaluated by taking all of the cases into account.

3. Relationship to Testability. Now consider the extent to which a theory can be confirmed if it is correct or falsified if it is wrong. A theory of the kind we are considering makes a number of empirical commitments regarding the validity of the domain and the properties that it specifies. However, the discussion here will involve only the assumed probability distribution, and for simplicity, only hypotheses about the conditional probabilities p_{ij} will be considered. Assume that the probabilities of the explanans are known, or as often happens, are controlled by experimental manipulations. In that case the hypothesis consists of n vectors of conditional probabilities and these, together with the known values of p_i determine the values of p_j;

$$p_j = \sum_{i=1}^{n} p_i p_{ij}.$$

A set of observations that can confirm or falsify the theory includes observations from some or all of the sets S_i. Let N_i be the number of observations from S_i, and assume that the observations constitute a random sample and that the size of each set S_i is large relative to N_i. The N_i observations will be distributed in some way among the m explananda, the M_j, so there will be m sample proportions $\hat{p}_{ij}$. If the theory is correct, the observed frequencies of the M_j will have the multinomial distribution, where for each j and j',

$$E(\hat{p}_{ij}) = p_{ij},$$

$$\text{Var}\,(\hat{p}_{ij}) = \frac{p_{ij}(1 - p_{ij})}{N_i},$$

$$\text{Cov}\,(\hat{p}_{ij}, \hat{p}_{ij'}) = \frac{-p_{ij}p_{ij'}}{N_i}.$$

The important property for evaluating hypotheses is the variance. If Var $(\hat{p}_{ij})$ is large, then the observations give very imprecise knowledge about the p_{ij}; the observed proportions $\hat{p}_{ij}$ may be quite different from the values specified by a theory but still not permit rejection of the theory, or the observed proportions may be quite close to those specified by the theory but provide only weak confirmation of the theory. There are standard procedures used for making statistical inferences in situations of this kind, but their details are not needed for the present discussion.

It is sufficient to note that the power of a set of observations in permitting inferences about a theory tends to be small when the variances of the sample proportions are large.

As an approximate index of the ability of a theory to be confirmed or falsified by a set of observations, consider a weighted sum of the variances of the $\hat{p}_{ij}$ values. We have

$$(14) \qquad \sigma = \sum_{i=1}^{n} p_i \sum_{j=1}^{m} \text{Var}\,(\hat{p}_{ij}) = \sum_{i=1}^{n} p_i \sum_{j=1}^{m} \frac{p_{ij}(1 - p_{ij})}{N_i}.$$

If σ is small the observations tend to permit stronger inferences than if σ is large.

It should not be surprising that the information transmitted is closely related to the variances of the sample proportions. When I_T is large, the conditional probabilities tend to be near one or zero, where the value of $p(1 - p)$ is small. This informal observation can be made slightly more specific. Using the approximation of equation (12), we have

$$(15) \qquad H(M) - I_T \approx \sum_{i=1}^{n} p_i \sum_{j=1}^{m} p_{ij}(1 - p_{ij}),$$

and the term on the right side is closely related to σ. For example, consider the case where the N_i are all the same so $N_i = N$, a constant. Then instead of equation (14), we have

$$(16) \qquad \sigma = \frac{1}{N} \sum_{i=1}^{n} p_i \sum_{j=1}^{m} p_{ij}(1 - p_{ij}) \approx \frac{1}{N}(H(M) - I_T).$$

Equation (16) shows that σ has a tendency to decrease as I_T increases. Given two theories that agree in the marginal probabilities of the explananda (the p_j) the theory with the higher value of I_T probably will be more testable.

Equation (16) also shows that testability is influenced by $H(M)$, and larger amounts of uncertainty in the explanandum tend to reduce testability. This fact relates to a comment made in Section **1** regarding testability of hypotheses about rare events. If $H(M)$ is low, there is less opportunity to reduce uncertainty about the system than if $H(M)$ is high. However, we now see that if $H(M)$ is low, hypotheses tend to be strongly testable even though I_T is small. The advantage of greater testability may outweigh the disadvantage of less information transmitted in some cases, especially when rare phenomena are important either practically or for theoretical questions.

4. Development of New Hypotheses. In most statistical theories, we lack nomological-deductive explanatory power because our knowledge is incomplete. A natural move is to try to obtain increased explanatory power by using a more detailed explanans. If the variables used to explain events permit only statistical explanation, this often is because additional variables are relevant but their relevance has not been discovered. The next result shows that the addition of a new explanatory variable will increase the information transmitted by a theory unless a very special condition of independence is satisfied. However, it also shows that the importance of a variable cannot be evaluated just be finding its relationship with the explanandum; the appropriate question seems to be how much explanatory power is added by the new variable, and this can only be ascertained by examining the way in which the new variable relates to all the variables in the theory.

We consider a statistical theory with explanans and explanandum

$$\{Q\} = \{Q_1, \ldots, Q_g, \ldots, Q_{n_Q}\},$$
$$\{M\} = \{M_1, \ldots, M_j, \ldots, M_m\}.$$

Then the information transmitted is

$$I_T(Q) = H(M) + H(Q) - H(Q \times M)$$
$$= \sum_{j=1}^{m} -p_j \log p_j + \sum_{g=1}^{n_Q} -p_g \log p_g - \sum_{g=1}^{n_Q} \sum_{j=1}^{m} -p_g p_{gj} \log p_g p_{gj}.$$

Now suppose that we add a new variable R to the explanans. The values of R are $R_1, \ldots, R_h, \ldots, R_{n_R}$, the new explanans is

$$\{S\} = \{Q\} \times \{R\},$$

a single value of S is

$$S_i = Q_g \cap R_h,$$

and the number of values of S is

$$n = n_q n_r.$$

The information transmitted by the new theory is

$$I_T(S) = H(M) + H(S) - H(S \times M).$$

We have

$$H(S) = \sum_{i=1}^{n} -p_i \log p_i; \qquad p_i = p_g p_{gh}.$$

Then

$$H(S) = \sum_{g=1}^{n_Q} \sum_{h=1}^{n_R} -p_g p_{gh}(\log p_g + \log p_{gh})$$

$$= H(Q) + \sum_{g=1}^{n_Q} \sum_{h=1}^{n_R} -P(Q_g \cap R_h)[\log P(Q_g \cap R_h) - \log P(Q_g)].$$

The uncertainty of the system is

$$H(S \times M) = \sum_{i=1}^{n} \sum_{j=1}^{m} -p_i p_{ij} \log p_i p_{ij}$$

$$= \sum_{g=1}^{n_Q} \sum_{h=1}^{n_R} \sum_{j=1}^{m} -P(Q_g \cap R_h \cap M_j) \log P(Q_g \cap R_h \cap M_j)$$

$$= H(Q \times M) + \sum_{g=1}^{n_Q} \sum_{h=1}^{n_R} \sum_{j=1}^{m} -P(Q_g \cap R_h \cap M_j) \\ \log P(Q_g \cap R_h \cap M_j) - \log P(Q_g \cap M_j)$$

$$= H(Q \times M) + \sum_{g=1}^{n_Q} \sum_{h=1}^{n_R} -P(Q_g \cap R_h) \\ \times [\log P(Q_g \cap R_h) - \log P(Q_g)]$$

$$+ \sum_{g=1}^{n_Q} \sum_{h=1}^{n_R} P(Q_g \cap R_h) \\ \times \sum_{j=1}^{m} -P(M_j \mid Q_g \cap R_h) \log \left[\frac{P(M_j \mid Q_g \cap R_h)}{P(M_j \mid Q_g)}\right].$$

Combining these results, we obtain

$$I_T(S) = I_T(Q) + \sum_{g=1}^{n_Q} \sum_{h=1}^{n_R} P(Q_g \cap R_h)$$

$$\times \sum_{j=1}^{m} P(M_j \mid Q_g \cap R_h) \log \left[\frac{P(M_j \mid Q_g \cap R_h)}{P(M_j \mid Q_g)} \right]. \tag{17}$$

The second term is always greater than or equal to zero. It is zero when for all g, h, and j,

$$P(M_j \mid Q_g \cap R_h) = P(M_j \mid Q_g).$$

In other words, the addition of a new variable to the explanans fails to increase the information transmitted when the conditional probabilities of M_j given all the variables equal the conditional probabilities of M_j given the old variables. In this case, we can say that $\{Q\}$ is a sufficient partitioning of Ω for the explananda $\{M\}$ [3]. Another way to see this is provided by Salmon's [4] notion of homogeneous classes. $\{R\}$ does not increase I_T if the sets of events described by $\{Q\}$ are homogeneous with respect to the explananda. It is easy to show that this is not equivalent to having

$$I_T(R) = 0.$$

That is, R by itself can be useful for explaining M but still provide no new information about M when it is added to an existing theory.

The above discussion involves finding a new variable to provide better explanation of a fixed explanandum. The same general properties apply when a fixed explanans is used to explain a more detailed explanandum. Suppose we start with a theory with explanans and explanandum

$$\{S\} = \{S_1, \ldots, S_g, \ldots, S_n\},$$

$$\{K\} = \{K_1, \ldots, K_h, \ldots, K_{m_K}\}.$$

The amount of information transmitted is $I_T(K)$. Now, we make the explanandum more detailed by adding a variable L to the description of events to be explained.

$$\{L\} = \{L_1, \ldots, L_i, \ldots, L_{m_L}\},$$

the new explanandum is

$$\{M\} = \{K\} \times \{L\},$$

and the number of values M takes is

$$m = m_K m_L.$$

By an argument exactly like that leading to (17), we can show that the information transmitted by the new theory is

$$I_T(M) = I_T(K) + \sum_{h=1}^{m_K} \sum_{i=1}^{m_L} P(K_h \cap L_i) \tag{18}$$

$$\times \sum_{g=1}^{n} P(S_g \mid K_h \cap L_i) \log \left[\frac{P(S_g \mid K_h \cap L_i)}{P(S_g \mid K_h)} \right].$$

Again, the usefulness of the new variable depends on its relationship with all the variables in the old theory. The new explanandum does not add any information transmitted if $\{K\}$ is a sufficient partition of Ω for the explanans $\{S\}$ and merely showing that $I_T(L) > 0$ does not demonstrate that L adds any new information.

The symmetry between equations (17) and (18) clarifies the relationship between prediction and explanation further. From a practical point of view, the goal of scientific work may be just to obtain as good an explanation as we can about a fixed set of explananda. For example, in engineering research about education, it often is sufficient to study the variables that influence achievement on a fixed set of tests. However, from the point of view of general scientific interest, these explananda may not be sufficiently detailed to provide a satisfactory theory of the processes of learning and motivation that are involved in education. If our goal is to increase our general understanding of a process, it may be just as useful to obtain more detailed descriptions of outcome variables (the explananda) and relate these to the known explanans, than to add to the variables that presently serve as explanans.

The discussion of the effects of adding variables to a theory seems to have implications for scientific practice. In many cases, it would be reasonable to expect a scientist to try to provide as much increase in information transmitted as he could. In that case, he should investigate hypotheses which by equation (17) or (18) would produce large values of I_T for the theory in case his hypotheses were confirmed. One implication seems quite clear. If the new variables to be investigated are closely correlated with those that already are included in the theory, they are unlikely to provide very much increase in information transmitted. The maximum of I_T is the uncertainty of the explanans or of the explananda, whichever is smaller. If we start with explanans $\{Q\}$ and investigate a new explanans $\{S\} = \{Q\} \times \{R\}$, we have

$$H(S) = H(Q) + \sum_{g=1}^{n_Q} p_g \sum_{h=1}^{n_R} -p_{gh} \log p_{gh},$$

which equals $H(Q)$ if Q and R are perfectly correlated and equals $H(Q) + H(R)$ if Q and R are independent. A similar relationship holds for additional detail in the explananda. If we start with explananda $\{K\}$ and investigate $\{M\} = \{K\} \times \{L\}$, we have

$$H(M) = H(K) + \sum_{h=1}^{m_K} \sum_{i=1}^{m_L} -p_{hi} \log p_{hi},$$

which is small if K and L are strongly correlated and large if K and L are independent. The guideline is a rule of thumb, of course. Two variables that are strongly correlated with each other may have quite different relationships with a third variable. But in many cases, it will turn out to be a good bet that if new variables are strongly correlated with old variables, investigation of those particular new variables will not lead to very large increases in information transmitted and therefore will not pay off very highly in terms of new understanding.

Partly for these reasons having to do with maximum information transmitted, and partly because of the relationship between I_T and testability, we have a situation here that fits with an earlier analysis [1] of the role of decision factors in scientific investigation. On the average, there probably is a negative relationship between the amount of increase in information transmitted due to a new hypothesis and the subjective probability of that hypothesis' being confirmed in an investigation. If we accept the present argument, then the value of confirming a new hypothesis should be positively correlated, on the average, with the increase in I_T that it yields. Then the investigator has to trade off a lower subjective probability of confirmation against a higher potential payoff in increased knowledge in selecting among the various hypotheses that he could investigate.

One final remark should be made regarding the development of theories of the kind analyzed here. We have noted that increased information transmitted almost always results when new variables are added to the theory. This almost suggests an algorithm for scientific investigation: increase the information transmitted by adding new variables. If it were not for the value of theoretical simplicity, this might be a good rule to follow. But simplicity is a virtue, as is obvious if we contemplate the development of theories toward infinitely long lists of variables and their intercorrelations. Happily, this kind of development can be complemented by development of hypotheses that simplify the theory. We can find that several variables really represent a smaller number of underlying theoretical processes. The theory about these processes may still involve statistical hypotheses, but the relationships among the theoretical variables may be considerably simpler than the theory that deals with observable variables or with theoretical variables that are less abstract. Examples of simplifying models are common in mathematical economics and psychology, and although a thorough analysis of their properties seems promising in connection with the present discussion, it will not be undertaken in this paper.

5. Conclusion. According to many analyses, an important characteristic of scientific theories is that they permit covering-law explanations of events, based on empirically testable assumptions. Analyses based on theories that permit nomological-deductive explanation have emphasized explanations of single occurrences. There have been problems in applying this idea to theories that include statistical hypotheses, and the emphasis in this paper has been on the overall explanatory power of the theory rather than on explanations of single occurrences. In this paper the information transmitted has been used as an index of a theory's explanatory power. This index has some properties that make it seem promising for evaluating statistical theories. Of course, any index of this sort will give only an approximate measure of a theory's power. There may be other ways of evaluating explanatory power that are either more accurate or more convenient. But the main argument of this paper is that an evaluation of the overall explanatory power of a theory is less problematic and more relevant as an assessment of the state of knowledge than evaluation of statistical explanations of single occurrences in terms of likelihoods that are assigned to explananda.

REFERENCES

[1] Emmerich, D. S., and Greeno, J. G., "Some Decision Factors in Scientific Investigation," *Philosophy of Science,* vol. 33, 1966, pp. 262–70.
[2] Hempel, C. G. *Aspects of Scientific Explanation.* New York: Free Press, 1965.
[3] Kullback, S. *Information Theory and Statistics.* New York: John Wiley, 1959.
[4] Salmon W. C. "Deductive and Inductive Explanation." Duplicated paper. August 1964. (The preliminary version of "Statistical Explanation," printed herein.)
[5] Scriven, M. "Explanations, Predictions, and Laws." In *Minnesota Studies in the Philosophy of Science,* vol. III. Edited by H. Feigl and G. Maxwell. Minneapolis: University of Minnesota Press, 1962, pp. 170–230.

Postscript 1971

The essay "Statistical Explanation" is reprinted here with only one significant change aside from the correction of minor typographical errors; namely, the inequality has been added to the formula on page 55 which defines the screening-off relation. Its omission from the original printing was a foolish oversight, and the need for the correction was called to my attention by J. A. Coffa. In order for D to screen off C from B, C must be irrelevant to B in the presence of D, but D must not be irrelevant to B in the presence of C. Coffa has also pointed out, quite correctly, that the *screening-off rule* enunciated on the same page is a derived rule—that is, this rule follows from the formal characterization of statistical explanation offered in section 13. The philosophical import of this rule seems sufficient, however, to justify its explicit recognition.

In section 6 I suggest that the explanatory question which calls forth an explanation contains a specification of a general reference class to which a prior probability (weight) can be referred. Professor Hempel rightly objected that the explanatory question may on many occasions contain no mention of any such class—for example, "Why was Caesar assassinated?"—and in some cases it may involve an inappropriate reference class—for example, "Why does this ice *cube* melt in my hand?" In the latter example the italicized term certainly introduces an irrelevancy which we shall want to exclude from the final explanatory account. Examples of both types strongly suggest that the first step in providing an adequate explanation may involve a reformulation of the explanatory question in order to introduce an appropriate reference class for the prior probability (weight) of the explanandum; thus we might ask, "Why was this supposed tyrant assassinated?" or, "Why does this piece of ice melt in my hand?" It was surely a mistake to have suggested that the uncriticized request for an explanation always points explicitly to the appropriate reference class for the prior probability (weight).

Hempel (and independently, Henry Kyburg) also raised a fundamental problem with my definition of "homogeneity" of a reference class as given on page 43: "If every property that determines a place selection is statistically irrelevant to *B* in *A*, I shall say that *A* is a *homogeneous reference class* for *B*." The question is whether, except in cases such as those in which every *A* is *B*, we can ever have a genuinely homogeneous reference class. For, the argument goes, any infinite sequence of *A*'s containing infinitely many *B*'s and infinitely many *non-B*'s has an infinite subsequence—say precisely the sequence of *B*'s—in which the relative frequency of *B* differs from the relative frequency of *B* in the entire sequence. If we assign the digit 1 to each occurrence of *B* in the original sequence *A* and the digit 0 to each *non-B*, the resulting infinite sequence of binary digits can be taken to define a real number between zero and one. Never mind how we know what number that is—the number exists and it defines a place selection that yields a relative frequency of *B* different from the frequency in the whole sequence.

This problem was posed long ago as an objection to von Mises's definition of a *collective*, and an answer was offered along the following lines: [1] Any normal mathematical language contains at most a denumerable infinity of expressions, so it has the capability of characterizing at most a denumerable subset of the nondenumerable infinity of real numbers between zero and one. Hence, although the number corresponding to any subsequence of *A* exists, only a denumerable subset of these numbers can be used to characterize place selections, for only a denumerable set can be incorporated into rules that can be expressed in the language. This argument was invoked by von Mises to prove that the concept of the collective is not self-contradictory and that collectives can exist; I would use it in the same way to guarantee the existence of homogeneous reference classes in nontrivial cases. I am prepared to admit, however, that there may be deep difficulties with the concept of randomness—and the associated notion of homogeneity—that are in need of further serious study.

Conversations with Coffa, Ronald Giere, and others have convinced me that some additional remarks should be made about *epistemic homogeneity.* If we have arrived at a reference class *A* in which we can make no further partitions relevant to the occurrence of *B*, there are two possibilities: we may have reason to believe that there are additional factors which we have not yet succeeded in identifying, or we may have reason to believe that there are no additional relevant factors in terms of which to make a partition of *A*. The paresis example illustrates the first possibility; in this case we do not know how to make a relevant partition of the class of untreated latent syphilitics, but we have good reason to believe that such a partition is possible in principle and that further medical research will likely reveal it. In this case we say that the class *A* is epistemically homogeneous. This judgment might be mistaken, but it is the plausible thing to say. The other alternative is illustrated by the example of spontaneous radioactive decay. Here we have good reasons for saying that the class of U^{238} atoms is homogeneous and that there exists no further relevant subdivision. This judgment might, of course, be incorrect—many able physicists believe that it is a mistake—but if it is wrong, then we must say that we thought the reference class to be homogeneous but in fact it turned out to be inhomogeneous. We should not say that the class was epistemically homogeneous, for to be epistemically homogeneous it must be *regarded as* relevantly subdividable, and the class of U^{238} atoms is not regarded as relevantly subdividable.

When Hempel and Oppenheim first offered the precise account of deductive-nomological explanation, they imposed the requirement that the statements contained in the explanans be true. Later, Hempel expressed a willingness to settle for statements that were highly confirmed. If we accept the earlier and more stringent requirement, we do not, of course, offer statements as explanatory unless we have good reason to believe them to be true—that is, unless they are well confirmed. We can never be absolutely certain that the statements we incorporate into an explanans are true; nevertheless, if we have good reason to believe them to be true, they can be utilized in proffered explanations. If we later find out that they are false, we should say that what we took to be an explanation was not actually an explanation, even though we had good reasons for thinking that it was an explanation. If, however, we accept the requirement that the explanans statements must be highly confirmed in terms of the total evidence available at the time the explanation is advanced, then we say that the explanation in question was a genuine explanation at the time, but now, in the light of additional evidence, a different explanation is required.

These two approaches are quite distinct. On the former approach, where truth is the requirement, there is such a thing as a genuine explanation objectively and quite apart from our knowledge situation. To be sure, the state of our knowledge determines *what we take to be* explanations, but no mention is made of our knowledge situation in the criteria for genuine explanations. On the latter approach, where high confirmation is the requirement, the very concept of a genuine explanation involves a reference to our knowledge situation, and there is no such thing as a correct explanation simpliciter.

If we acknowledge the possibility of explanations in both senses—relativized and nonrelativized—it would seem to be a matter mainly of linguistic convention how we decide to use the word "explanation." However, Hempel has argued repeatedly and at length that inductive-statistical explanation *must* be relativized to the knowledge situation, and this would seem to be a thesis of considerable philosophical import.[2] The statistical-relevance model, in contrast, allows statistical explanation in a nonrelativized sense. To have this type of explanation, it is necessary to partition the original reference class into actually homogeneous subclasses. If we define S-R explanation in terms of epistemic homogeneity—that is, the reference classes in the final partition are either actually or epistemically homogeneous—then we have relativized S-R explanation to a knowledge situation. Thus, according to the S-R account, one is free to choose between a relativized and a nonrelativized account of statistical explanation, whereas on Hempel's view, the relativized account is mandatory. This basic difference between the two approaches seems to rest upon the fact that Hempel appeals to the logical concept of probability with its requirement of total evidence, whereas my theory is given entirely in terms of objective homogeneity within the frequency interpretation of probability. This difference is further signaled by the fact that the frequency interpretation utilizes an objective concept of randomness, whereas the logical interpretation relativizes randomness to the total knowledge situation.

Throughout "Statistical Explanation," I have confined attention almost entirely to the explanation of particular events, and in section 14 I have disclaimed any attempt to treat theoretical explanation. The fact that a number of examples were taken from such highly theoretical contexts as nuclear physics does not really alter the situation, for these examples were chosen to provide the most convincing illustrations of occurrences that can with good reason be regarded as irreducibly statistical. In particular, theoretical explanations of the "observed" statistical regularities invoked in the explanations of individual events were not under discussion. Now it may be that theoretical explanation involves nothing but the subsumption of lower-level generalizations under higher-level generalizations (universal or statistical), but I have not tried to argue that this is the case, nor do I think it very likely to be true. Theoretical explanation looks very much as if it involves considerations beyond those involved in the explanation of particular events by subsumption under general laws.

As I said in the beginning of the Introduction, scientific explanations are sought for practical as well as intellectual reasons. In many instances, as William P. Alston argues in a recent article,[3] the search for explanations of particular events is motivated by practical concerns: the FAA seeks the explanation of a particular airplane accident, the psychiatrist seeks an explanation for a certain individual's neurotic symptom, and the engineer seeks an explanation for the collapse of a particular bridge. These all seem to be cases in which the general laws are not in question, but the problem concerns the initial conditions applicable to the particular event. Evidently, the search is for a cause, the discovery of which will be useful in that particular case or others quite like it. There is a rather clear sense in which explanations of this sort are of interest to applied, as contrasted with pure, science.

There may be deep significance to the distinction between the attempt to explain particular events and the attempt to explain general regularities (even though the distinction itself may be very difficult to characterize). For statistical regularities at least, this difference corresponds to the difference between the limiting frequency, which is a genuine probability, and the weight attached to a single case, which is a "fictitious" probability assigned to the single event for practical rather than theoretical reasons. As I have suggested in discussing the problem of the single case, the limiting frequencies are the entities with which probability theory deals, whereas the assignment of weights to single events amounts to the choice of a number useful, for example, as a betting quotient in situations where practical decisions are in order.

There are, however, other circumstances in which the scientist has a more purely intellectual interest in explanations of particular events, namely, when the particular occurrence appears to be in conflict with theoretical predictions. One can well imagine Michelson, who was attempting a precise determination of the velocity of the earth relative to the ether, asking in exasperation why the interference fringes did not shift when the apparatus was rotated. But it can easily be seen that he was really concerned with the general question of why interference-fringe shifts do not occur when a particular *type* of experiment is undertaken. If the fringe shift had only failed on one particular try, but occurred quite regularly on other trials, it is quite possible that the experimenter would have brushed aside the one failure on grounds (possibly uninvestigated) of equipment failure. As Karl Popper has pointed out, a refuting instance of a theory does not refute unless it is repeatable.

Theoretical science is obviously concerned with generality. While the FAA investigator, with his practical concerns, may be satisfied to *know* the general inverse relation between density and moisture content of air, the beginning of genuine intellectual understanding seems to come with the *explanation* of that regularity in terms of such theoretical considerations as Avogadro's law and the molecular constitution of the air. Similarly, although the nuclear engineer may be satisfied with a knowledge of the half-lives of various isotopes, the theoretical scientist wants to understand spontaneous radioactive decay in terms of the tunnel effect and the general quantum mechanical wave equation that predicts and explains the observed frequencies of radioactive disintegration. It is unnecessary to belabor this point with many examples. Although explanation of particular events has its place (usually in applied science), scientific comprehension seems to demand theoretical explanations of general regularities in nature, and our intellectual curiosity cannot be satisfied without them. It appears that the most pressing philosophical problems concerning scientific explanation center upon the nature of theoretical explanation.

NOTES

1. See Richard von Mises, *Probability, Statistics and Truth,* 2d Eng. ed. (London: George Allen and Unwin Ltd., 1957), pp. 92–93.
2. Carl G. Hempel, *Aspects of Scientific Explanation* (New York: Free Press, 1965), p. 402.
3. "The Place of Explanation of Particular Facts in Science," *Philosophy of Science,* XXXVIII, 1 (March 1971), pp. 13–34.

Bibliography

The published literature on the topic of scientific explanation is extremely voluminous; the following six books are standard works:

Braithwaite, Richard Bevan. *Scientific Explanation.* New York and London: Cambridge University Press, 1953. Reprinted as a Harper Torchbook, 1960.

Feigl, Herbert, and Maxwell, Grover, eds. *Scientific Explanation, Space, and Time.* Minnesota Studies in the Philosophy of Science, vol. 3. Minneapolis: University of Minnesota Press, 1962. Essays by Paul K. Feyerabend, Carl G. Hempel, Michael Scriven, and May Brodbeck.

Hempel, Carl G. *Aspects of Scientific Explanation and Other Essays in the Philosophy of Science.* New York: Free Press, 1965.

Nagel, Ernest. *The Structure of Science: Problems in the Logic of Scientific Explanation.* New York: Harcourt, Brace & World, 1961.

Rescher, Nicholas. *Scientific Explanation.* New York: Free Press, 1970.

Scheffler, Israel. *The Anatomy of Inquiry.* Part I. New York: Alfred A. Knopf, 1963.

Rescher's book contains a comprehensive, up-to-date bibliography. Two current items, not mentioned by Rescher, bearing directly on the S-R model of explanation are:

Coffa, J. A. "Foundations of Inductive Explanation." Ph.D. dissertation, University of Pittsburgh, 1971(?).

Greeno, James G. "Theoretical Entities in Statistical Explanation." In *Boston Studies in the Philosophy of Science,* vol. VIII. Edited by Robert S. Cohen and Roger C. Buck. Dordrecht-Holland: D. Reidel Publishing Company, 1971. With comments by Richard C. Jeffrey and Wesley C. Salmon.

Index